From *Lattu* to Lasers:
Realising India's Electricity Potential

About the series 'Books for the Concerned Citizen'

Leveraging the diverse expertise of its members in the subject domains and in publishing, the TERI Alumni Association is publishing a series of books on topics related to energy, resources, and the environment. The idea is to share information and, even more important, critical insights and understanding, with citizens who are keen to know more about some of the critical issues facing society and the world today but are lost in the deluge of information.

Our target audience is educated adults who are concerned about topical issues but lack the understanding to make sense of what they read or watch in the mass media—the series aims to equip them with conceptual tools and essential information not only to enrich their understanding but also to encourage them to act and thereby, albeit indirectly, further the UN Sustainable Development Goals.

The topics to be covered in the series and their respective subject-matter-specialist authors are listed below.

- **Rooftop solar***: Suneel Deambi and Shirish Garud
- **Coal***: Rakesh Kacker
- **Sustainable buildings***: Mili Majumdar and Minni Sastry
- **Nutraceuticals***: Mayurika Goel
- **Electricity**: Sanjeev S Ahluwalia
- **Public transport**: Shri Prakash and Sharif Qamar
- **Energy efficiency**: Ajay Mathur and Lehcr Thadani
- **Climate change**: Manish Shrivastava

already published and available for purchase

The publication of this series is financially supported by the Shakti Sustainable Energy Foundation. All books, being printed and marketed by TERI will be published latest by October 2022.

From *Lattu*s to Lasers
Realising India's Electricity Potential

Sanjeev S Ahluwalia

© TERI Alumni Association 2022

ISBN: 978-93-94657-02-1

Suggested citation
Ahluwalia S S. 2022. *From* Lattu*s to Lasers: Realising India's Electricity Potential.* New Delhi: TERI Alumni Association. 72 pp.

TERI Alumni Association
Administrative Wing, TERI
Darbari Seth Block (ground floor)
Habitat Place
Lodhi Road
New Delhi – 110 003

> Price Rs 299/-
> For sales queries, please contact us at
> Nand K Yadav, Assistant Manager - Sales
> The Energy and Resources Institute (TERI Press)
> Darbari Seth Block, Habitat Place
> Lodhi Road, New Delhi – 110 003
>
> +91 97 173 56537 or +91 (0) 11 7110 2100 or 2468 2100
> nandkumar.yadav@teri.res.in or teripress@teri.res.in
> Fax +91 2468 2144 or 2468 2145

For more information, contact
Sanjeev S Ahluwalia (ahluss@gmail.com)

To

Professor Surendra L Rao
founding chairman of the CERC
mentor, friend and collaborator extraordinaire

CONTENTS

Foreword	ix
Preface	xi
Introduction	1
Electricity dispels the darkness	2
The origins of electricity	3
The colonial period: electricity for the elite: 1950–1984	4
Democratizing electricity supply	9
The Electricity (Supply) Act, 1948	10
Central planning favours public investment-led development	12
The Industrial Policy Resolution, 1956	13
The fiscal cost of public electricity supply	16
Electricity-intensive development	16
Changes in consumption pattern	18
The return of private investment: 1985–2020	22
International comparison of electricity supply	23
Profligate use of electricity shunned	25
Incentives for electricity reform	25
Union government's initiatives for structural reform in electricity	26
The outcomes of reforms	36
The end of supply shortages	36
Generation capacity utilization lower than optimal	36
Transmission shines	37
Private power exchanges	38
Discoms: the weakest link in the value chain	38
The unfinished reforms agenda	39

Planned electricity development: supply-side triumph or copycat industrial policy? — 40

- Putting industrial development above basic social and human needs — 40
- Throwing out the baby with the bathwater — 41
- Misallocation of public funds — 43
- Borrowed templates — 45
- Path dependency — 46
- Economic growth and competitiveness — 47
- New pathologies — 48

Unresolved issues in electricity supply — 49

- Electricity suppliers: too few or too many? — 49
- Has competition in supply increased? — 51
- Has the unbundling splintered supply to unviable levels? — 52
- Measures to enhance competition — 52
- Autonomy for regulators — 56
- The needs of viable power markets — 63
- The need for a 'smart' grid — 64
- Four short-term hard choices — 64

Trends favouring India — 66

- High growth can make green power affordable — 67
- Renewable and hydroelectricity offer unutilized potential — 67
- Growth of digital connectivity — 67
- Symmetric policy preferences across parties — 67

Bibliography and notes — 69

FOREWORD

I am happy to see this concise volume by Sanjeev Ahluwalia. I had the good fortune of engaging with Sanjeev when he was the 1st Secretary to the Central Electricity Regulatory Commission in 1999. This first Commission adopted bold procedural measures to ensure transparency and consumer participation in the functioning of the Commission. These Conduct of Business regulations were adopted by many state commissions and were instrumental in ensuring transparency and consumer participation in these newly formed institutions across the country. It is a different story that over the last couple of decades some state commissions have diluted these provisions and, in the process, eroded their credibility.

This compact volume crisply traces the history and evolution of the electricity sector in India over one and half centuries. It lucidly narrates various legal and policy frameworks that guided the sector all these years. Alongside this historical journey, he also contextualizes many challenges that the Indian electricity sector is facing. He highlights how electricity is an essential input for development and how India has a long way to go to meet its electricity needs for a decent standard of living for all its residents.

World over, and also in India, the electricity sector is undergoing turbulent times like other sectors of the economy. This sector, which contributes most to the greenhouse gas emissions and climate change, has to deal with the challenge of rapidly decarbonising electricity generation and also facilitating decarbonization of other sectors by electrifying many other end-uses such as transport. In addition to this, the Indian electricity sector, which was already facing financial distress and viability challenge, faced a big blow due to the COVID pandemic. As if this was not enough, the Ukraine crisis demonstrated the vulnerability of the sector to the global supply chain and price shocks.

The book does not stop at only presenting the history and challenges before the sector but also proposes certain bold reform measures, including privatization of distribution and establishment of regional regulators. The electricity sector is complex, and any reform measure needs to be thoroughly discussed considering the natural monopoly nature of the wires portion, changing competitive supply scenarios due to falling costs and modular nature of renewables, governance experience and capability, past experiences, as well as reforms needed in the fuel sector.

There may be divergent opinions about some of the measures the author suggests, but I am sure every stakeholder in the sector will certainly agree with the following observation of the author: "The third decade of this century has dawned with reminders that India needs to enhance its energy resilience and security. This is possible if the spirit of cooperative federalism suffuses institutional arrangements and public interest trumps political upmanship."

This book is a valuable addition to the sector discourse in this context and I compliment Sanjeev, TERI Alumni Association and SSEF for bringing out this volume.

August 2022

Shantanu Dixit
Prayas (Energy Group), Pune

PREFACE

It is daunting to author a book on Indian electricity for the casual reader because of the sheer volume of written work which already exists and the daily deluge of information in the media. A consequential question is, why then even attempt to do so and what have I added that readers did not already know? This book is based on secondary sources. However, that still leaves scope for evidenced opinions which I have been liberal in sharing.

Facts are facts, of course, but it is in the conjunction of one fact with another that interesting conclusions can be reached and mulled over. It is my hope that the reader will discover her area of interest in the book and use some of the arguments expressed here to reach her own conclusions.

The history of electricity in India traces an inverted arc – like a smiley – which starts in the colonial period with private electricity capacity leading, regresses to a mode of near complete public sector monopoly by the 1980s and then traces the upward incline to a near 50% share for private electricity suppliers – not a full smiley but a slightly lop-sided one. The half-smile – like Mona Lisa's – masks long periods of misallocation of public capital, unabashed populism, and careless adherence to 'path dependencies' which plagues bureaucracies the world over.

This book asks a few inconvenient questions and provides some out-of-the-box solutions with the intention of enlarging the public debate around how the electricity sector should be regulated and developed going forward.

This work was started in the waning months of the coronavirus epidemic at our home in village Dhamas, Almora. Whilst the weather was congenial, this put paid to any library research I might have done. I was delighted to discover the vast web resources available, which advanced my understanding of how it all began and why, in my view, things went seriously wrong three decades after Independence. By then 'central planning'-based public investments had reached unsustainable levels.

I am grateful to our pet, Sheru, who kept me company on the writing trail and to Vidyun, my wife who indulged me; Armaan, our son for keeping my spirits up; and Ishaan and Roma, our elder son and daughter-in-law, for diligently teleconnecting from Mumbai.

I owe a debt of gratitude to Shantanu Dixit of Prayas for reading an advance draft and agreeing to write the foreword. I have also benefited from the generosity, industry smarts and vision of Dr Pramod Deo, Chairman, CERC, 2008–13; Shri Gireesh Pradhan, Chairman, CERC, 2013–17; and Shri Prabir Neogi, electricity supply industry veteran and Chief Advisor (Corporate Affairs), RP-Sanjiv Goenka Group.

Grateful thanks also to Rakesh Kacker and Dr Vibha Dhawan, Vice Chairperson and Patron, respectively, of the TERI Alumni Association (TAA) for reposing their faith in me, to my wonderfully perceptive editors Yateendra Joshi and P K Jayanthan and Anupama Jauhary, Rajiv Sharma, and the team at TERI Press for their patience and technical excellence.

Finally, I acknowledge the generosity of Sunjoy Joshi, Chairman and Dr Samir Saran, President, Observer Research Foundation, in encouraging me to contribute to this series of Concerned Citizen publications by the TERI Alumni Association and Shakti Sustainable Energy Foundation.

Introduction

India first experienced *lattus* (electric bulbs which light up when connected to a generator) in Surat, a textile industrial city and port on India's western seaboard with a population of around 89,505 (1851 census) in 1883. Sadly, the experience was short lived as the entire installation was damaged by floods in the local river Tapi. But it sparked a journey to universal electricity access, which ended 139 years later when the night sky above Rashtrapati Bhawan in New Delhi was lit up by laser beams in a spectacular display of optical wizardry for the Beating the Retreat ceremony during the Republic Day celebrations in 2022.

Those who watched the show must have thought that India had stepped seamlessly from the past into the future. We do not know if the residents of Surat also felt the same sense of elation and pride, for they too had stepped into the future through a time window. We do know, however, that thereafter the darkness of the night was dispelled for millions of Indians.

This book recounts the discovery of electricity in the West and its relatively rapid spread to India, as an elite perquisite in the colonial period. After Independence, in 1947, the political compulsion to provide access to electricity for all, led to patchy but determined growth in supply. It took three-fourths of a century to achieve this objective: not surprising, given the huge effort required to electrify the second largest population in the world, constrained by lack of fiscal resources.

The path adopted to enlarge access to electricity during 1950–1984 was through planned development and primary reliance on the public sector. Outcomes were better in providing access than in improving the quality of supply or financial stability of the utilities. During 1985–2020, an opportunity arose to adapt to changing global circumstances and explore liberalized options bringing in private sector supply to augment public sector capacity in generation. This reform included industry restructuring and the introduction of autonomous regulation.

Was the decision to use only public resources for electricity provision, taken in 1956 and continued till 1991, inevitable, or was it a costly mistake defining pathways which continue to constrain private enterprise in India? Will India recover from the low-level equilibrium of poor operational efficiency, excessive costs, low revenues, and high indebtedness, which characterize the publicly owned electricity retail supply segment?

Can autonomous regulation, commenced in 1998 with lofty expectations, shed its image of timidity and subservience to political compulsions? Are there options to strengthen the ability of the regulatory bodies to give primacy to economically beneficial and technically sound solutions to the unresolved issues in electricity supply? How can competition rationalise retail prices, enhance incentives for energy efficiency and promote green energy?

This book provides the context, shares the historical evidence, and suggests options. India has significant comparative advantages: high potential for economic growth, unutilised potential for domestic production of renewable energy, rapidly progressing digital connectivity and broad political consensus on economic policies for transitioning to a mature, stable, market oriented and resilient electricity supply system providing high quality, sustainable services to all.

Electricity dispels the darkness

Electricity followed the telegraph and the railways in finding a foothold in British India. The first telegraph line was built in 1851 between Calcutta and Diamond Harbour.[1] Between 1853 and 1870, keeping military and economic necessities in view, more than 6400 kilometres of railway lines were laid.

The speed of technology transfer was astonishing despite the extended lags of the time: a journey from Bombay (Mumbai) to

London took more than two months, and 'snail mail' within India was carried by horse- or camel-mounted couriers, pack animals, bullock carts or boats. The first railway line between Liverpool and Manchester in England[2] had become operational only two decades earlier, in 1830. The first telegraph system between Washington, DC and Baltimore, Maryland, funded by the US Congress, was set up and tested by Morse and Vail less than a decade earlier in 1843.[3]

The origins of electricity[4]

Humans are believed to have controlled and managed the use of fire about 40,000 years ago by using dung, a slow-burning fuel, to keep the hearth burning. They sourced fire from wildfires, started by lightning striking dry grass and trees. Scientists believe that it would have been impossible for humans to develop their brain power without eating cooked food. One happy outcome of the growth in brain size and thinking ability was their singular capacity, amongst all living beings, to control and manage fire for warmth and to cook.

Electricity, as a force of nature, has been visible to humans since millennia in the flashes of lightning followed by the startling thunderclaps on stormy nights. We have known since the mid-18th century, thanks to Benjamin Franklin in Philadelphia and subsequent collaborative experiments by scientists in Paris, that lightning is a giant electricity charge caused by the balancing movement of electrons between positively and negatively charged particles in clouds, the air, or on the ground.

By 1800 Alessandro Volt applied this scientific knowledge to develop the first battery using copper and zinc plates, separated by cardboard soaked in brine, to provide a low voltage but consistent current. Volt, a unit of electromotive force, or the difference in potential, which causes a current of one ampere (a unit to measure the flow of electricity) to flow through a resistance of one ohm (a measure of resistance), is named after this Italian physicist.

Kinetic energy, as in the force of river flow, was traditionally used to run grain mills (chakkis), but its application for generating electricity had to await the British scientist Michael Faraday when he discovered, in 1831, that moving a copper wire across a magnetic field produces electricity. This was followed by commercialization of generator technology using permanent magnets to light the arc lamps in lighthouses, which is credited to H Wide of Britain. By 1866 Werner von Siemens produced a dynamo generator with self-energizing magnets. In 1882 the first hydroelectric plant was commissioned in Appleton, Wisconsin, as backward integration after Thomas Edison had patented the incandescent light bulb in 1880. Soon electricity became a public utility, lighting the streets of major cities in America and Europe.

The colonial period: electricity for the elite: 1950–1984
Surat lights the way

In 1883, just five years after the streets of European cities were lit by arc lamps, the main street of Surat, in Gujarat, was similarly lit up by a small generator. Arc lamps were the predecessors of the more efficient incandescent bulbs, colloquially known as *lattu*s in north India after the similarly shaped spinning top, a favourite and cheap popular toy. Sadly, the entire installation was washed away by floods in the river Tapi soon after. However, by then the desire to dispel the night effortlessly and cleanly had spread across India.

The Darjeeling municipality installed a small hydroelectric generator of 130 kilowatts to power 200 streetlights on the main roads of the town in 1897, more than a decade after the first hydroelectric plant was set up in Appleton, Wisconsin, USA, by the Appleton Edison company in 1882.[5] Unsurprisingly, imperial Calcutta (now Kolkata), the then capital of the British Empire in India, was not to be left behind. Streetlights were installed there in 1892, followed by the promulgation of the Calcutta Electric Lighting Act in 1895 laying the basis for distributed electric supply to public and private spaces in the Bengal Presidency.[6] The Madras and Bombay presidencies followed, and electricity got a foothold in popular imagination as an aspirational symbol of convenience and modernity although only the privileged and the rich had access to electricity.

The ingenuity of private enterprise and the supportive actions of the colonial government in funding implementation, as in the case of telegraph lines or leasing of government land or offering private investors assured returns of 5% for investing in the railways, are examples of early public–private partnerships, a winning model of infrastructure development in the colonial period.

Light spreads to the Presidency towns and princely states
When electricity became available in Calcutta in 1902, it was priced at half a rupee per unit of electricity (kilowatt-hour), affordable by only the wealthiest homes: only 70,812 domestic customers out of a population of 933,754 in 1901.

The princely states of Hyderabad and Mysore were electrified even before Bombay, Delhi, and Bangalore. In 1901, Hyderabad became the first city to be electrified. Between 1932 and 1939, electricity consumption in Hyderabad state grew from 10 million units to 16 million units and the number of consumers went up from 4000 to 7500. This surge in the demand occurred because the tariff was reduced from half a rupee to a little over an eighth of a rupee per unit, which was a quarter of the rate that was charged when electrification started in Calcutta.

In the state of Mysore (now Karnataka), completion of the Shivanasamudra project on the Cauvery River led to the generation and supply of hydroelectricity in 1902. The government encouraged rural electrification by reducing the price of electric irrigation pumps from 200 rupees to 100 rupees. The tariff for small industries was also reduced to 1/16th of a rupee (1 anna) per unit by 1928. By 1936, there were 383 irrigation pump-sets, 158 flour mills, and 637 rice mills driven by electricity in Karnataka.

Electricity reached Delhi in 1902 to decorate the city for the 'Imperial Durbar' (to celebrate the coronation of Edward VIII) as well as to power trams. One of the first localities to be electrified after the Durbar was the civil lines. The Delhi Electric Tramways and Lighting Company built a power station just outside Lahori Gate.

In Bombay, the initial experiment with lighting in Crawford Market failed. The Bombay Electric Supply and Tramways Company was set up in 1905 and was licensed to supply electricity to the city. It started tramway services two years later and began supplying electricity to domestic and commercial consumers from a 4.3 MW plant set up in Wadi Bunder. As the tramway services expanded, another station was set up in Mazagaon in 1912.

Bangalore and Lahore (now in Pakistan) were electrified in 1911 and Jalandhar in Punjab, in 1925. The electrical department of Punjab reported that 10,458 streetlights, 3223 fans, 77 motors, 35 pump-sets, and 36 radiators were operational in the region in 1923/24.

A luxury unaffordable for most

But the high cost of electricity – an electric bulb cost 2 rupees – made it unaffordable for most households (Table 1), considering that the pay of an Indian sepoy – at the time an entry-level position of significant prestige and value – fighting for the British in the First World War was only 11 rupees a month: a modest monthly consumption of 25 kWh would have amounted to approximately 28% of the monthly wages, a fairly high expense for only basic lighting and perhaps a fan. A corresponding level of convenience assuming energy-efficient bulbs, a desert cooler, and a fan – amounting to a monthly consumption of 50 kWh – would amount to only 1.25% of an Indian soldier's salary of Rs 20,000 a month a century later in 2014.

Table **1** The affordability of electricity: 1920s versus 2014

Description	Year	Unit	Value
Pay of sepoy per month	1920s	annas	176
Cost of 1 kWh	1920s	annas	2
Cost of 25 kWh	1920s	annas	50
Electricity cost/pay per month		%	28.4
Pay of jawan per month	2014	INR	20000
Cost of 1 kWh	2014	INR	5
Cost of 50 kWh	2014	INR	250
Electricity cost/pay per month		%	1.25

SOURCES 1920s data "Steps of power: Notes on the history of electrification in India (1883–1930)". Proceedings of the Indian History Congress, Vol. 78. 2017. pp. 498–506 and "India and the Western Front" by David Omissi https://www.bbc.co.uk/history/worldwars.

Electricity has become much more affordable over time not only because of technological improvements and higher efficiency in its production and supply but also because of higher salaries in the formal sector of the economy (which includes the public sector), particularly for workers at the lower levels who are protected by minimum wages legislation and a progressive policy of higher wages based on equity considerations. This example illustrates the triple political incentive behind providing electricity as a public good. First, electricity drives economic growth; second, it enhances personal convenience; and third, scaling up the volume means the significant capital cost can be spread over a larger production level, reducing the per-unit cost of supply (called the economies of scale).

Regulating supply: The Electricity Act, 1910

The supply of electricity in selected areas across the Indian peninsula by early 20th century and the prospects of its wider use in the future induced the need for legislation to standardize its quality and supply.

The Electricity Act, 1910, applicable to all British territories in India, came into force on 18 March 1910 and formalized the procedures for the grant of licenses to supply energy. The act prohibited the transfer or sale of the license without approval; prohibited the association of two licensees without prior approval; and required annual submission of accounts by the licensee. The act also laid down procedures to be followed by licensees while laying new lines or repairing existing lines to minimize inconvenience to owners or occupants of the property through which the lines passed.

The act also authorized a licensee to enter the premises of customers for maintenance or repair, to inspect the supply connections, or to check the meter. The licensee was obliged to supply power on demand to a customer on condition that the customer pays the cost for such supply, including in cases where only emergency or temporary back-up supply might be required. An electricity inspector was empowered to enquire into cases of damage or loss of life due to malfunctioning of electrical supply lines. Penalties were imposed for actionable felonies such as tampering with the meter, stealing electricity, or damaging supply lines, public lamps, or other equipment of the licensees or for wasting energy.

While the regulations were comprehensive on the physical aspects of supply and safety, they did not explain how electricity tariffs should be determined or if differential terms should apply to extend access to electricity. This imbalance reflected the thinking of the times that the licensees should supply power on commercial terms as best suited to them. Electricity was viewed as a convenience for those who could afford it, rather than an input into improving the quality of life and enhancing economic development; if it was, such were not the primary objectives of the colonial administration at the time.

Electrifying India not a colonial priority

The focus of the colonial government in the infrastructure sector during the first quarter of the 20th century was on railways, roads, and irrigation, in that order, in terms of budgetary outlays, with electricity getting a small share of public investment only in the second quarter of the century (Table 2).

Table **2** Public spending on select infrastructure

Sector	1861–1919		1920–1947	
	Rupees (**million**)	% Share	Rupees (**million**)	% Share
Railways	3818	53.7	6168	58.3
Roads	1570	23.7	2237	21.2
Irrigation	1227	22.6	1968	18.6
Electricity	0	0	199	1.9
Total	6615	100	10572	100

SOURCE M J K Thavaraj in Tirthankar Roy, *The Economic History of India.*

During the first quarter of the 20th century, railways had the highest share, at 58%, among the four selected infrastructure sectors, followed by roads and irrigation, with nearly similar outlays. These allocations reflected the economic, strategic, and security concerns of the colonial government—the share of electricity was a paltry 2%.

The Indian Electricity Act, 1910 (with amendments) was in operation until 2003, when it was replaced by the Electricity Act, 2003.

Democratizing electricity supply

India attained Independence on 15 August 1947. The Planning Commission was established in 1950 and was to remain the central point for coordination of public investment across the union and state governments, until the commission was dissolved more than six decades later, in 2015.

At the commencement of planning in India in 1950, electric connections had been extended to fewer than 3061 villages.[7] This was inevitable because most of the capacity had been created by private enterprises and meeting the cost of supply through tariff was an important commercial consideration. The resultant inequity in access to electricity was unacceptable. The solution chosen was to side-line private investment and ramp up public investment in electricity.

This resonated with the changed architecture of governance post-World War II (1939–45). Intrusive public regulation, necessary to manage the privations of the war, got firmly embedded in the substance of government functioning including industrial licensing, controls over supply and consumption, price constraints, and a tendency towards 'big government', which was retained well after global peace had been restored.

In the UK, the Labour party, which came to power immediately after the end of the Second World War, embarked on a social welfare programme of nationalizing major industries including profitable ones such as steel, railways, and coal mining. Over 500 local private electricity companies were brought under a central entity, the British Electricity Authority, with 12 boards under it, one for each area.

Private electricity capacity in India was limited in comparison. India's per capita electricity consumption was only 15.3 kWh in 1947 compared to about 1300 kWh in Britain in 1948. The nationalization of the major industries in Britain and the growing economic strength of the Soviet Union provided compelling examples of 'big government', which a fledgling nation such as India would have found difficult to ignore.

The Electricity (Supply) Act, 1948

The Electricity (Supply) Act, 1948, was passed to provide a comprehensive framework for the efficient regulation and growth of electricity supply. It established CEA, the Central Electricity

Authority, under the union government, as (a) the technical arbiter between state governments, (b) the primary source for developing technical norms for generation, transmission, and distribution of electricity, (c) the agency for the formulation of procedures for safe and stable management of regional grids, and (d) a source of technical information, data, and research. The agency sought to integrate what until then had been separate generating and supply systems into a single networked grid system within a province and further integrated into regional grids to facilitate seamless electrification across urban and rural areas.

The credit for focusing on the need to develop and maintain a stable, coordinated grid within regions, the framework on which the pan national grid was built subsequently, linking customers and suppliers for providing least-cost electric supply goes to the Electricity (Supply) Act, 1948.

The act was modelled on the Electricity Supply Act 1926 of the United Kingdom and created local public monopolies (SEBs, or state electricity boards) for buying electricity from generators as well as for generating electricity and supplying it at regulated tariffs to individual and bulk consumers.

Practicality trumps ideology

In a welcome departure from the UK legislation, the Electricity (Supply) Act did not attempt to create a single centralized integrated generation and monopoly supplier like the British Electricity Generating Authority, nor did the act simultaneously nationalize the existing private licenses under the 1910 act. The Calcutta Electricity Supply Company (now CESC) is one such entity that continues to be an electricity generator and supplier: it has diversified to other states – Uttar Pradesh, Rajasthan, and Maharashtra – primarily in the distribution business.

Private generation and supply accounted for 65% of the total supply until 1947,[8] which explains the practicality of maintaining continuity in their operations despite the new state-led development proposed by the 1948 supply act. Nevertheless, the creation of SEBs as specialized agencies of state governments to act both as regulators within their jurisdictions – determining tariff for bulk and retail supply based on normative costs – and as generators and suppliers to customers, backed by state fiscal power, inevitably favoured growth of the public sector.

In 1951, there were over 300 private licensees and 270 state-government licensees. By 1976/77, the number had dwindled to only 49 private licensees and 21 municipal undertakings, of which only 12 were generating and distributing electricity. The state electricity boards, as they grew, absorbed many of the erstwhile private licensees taking the share of the public sector to 84%.[9]

Private investment in electricity loses its shine

The shift in policy to favour public ownership and public investment in electricity had unnerved private investors, a fear reflected in a speech by A N Haksar, chairman of BSES Limited, at the annual shareholders' meeting in August 1960:[10] "... it is the proclaimed policy of the Central Government that Electric Supply Undertakings rendering valuable service should not be displaced merely for the sake of displacement. I hope I am right in saying, therefore, that the investor in the Corporate Sector of the Electric Supply Industry need not worry too much about premature acquisition."

Central planning favours public investment-led development

The institutional impetus towards electricity development led by the public sector under the act of 1948 was strengthened by the onset of central planning. The five-year plans which commenced from 1950 mapped and forecast fiscal resources likely to be available to the union and state governments along with prioritized development needs and proposed fiscal allocations jointly from the union and state government budgets for specific projects and programmes spanning all the development heads of expenditure in the annual budget (excluding defence and security). The growth in the volume of planned investment in the electricity sector during 1951 to 1992 is shown in Table 3.

Table 3 Public investment and generation capacity added (1951–1979)

	Investment in **Rs** (current) billion	% Share plan expenditure	Gen. cap. added in **GW**
First Plan 1951–56	2.6	13	1.1
Second Plan 1957–61	4.5	9	2.3
Third Plan 1962–66	12.5	16	4.5
Annual Plans 1967–69	12.1	19	4.1
Fourth Plan 1970–74	29.3	18	4.2
Fifth Plan 1975–79	73.9	19	10.8
Annual Plan 1979–80	22.4	18	2
Sixth Plan 1981–85	182.9	19	16.4
Seventh Plan 1986–90	378.9	21	24
Annual Plan 1990–92	256.1	20	6.6
Total	975.2		76

SOURCE Annual Report on the working of SEBs and EDs. Planning Commission. 1995.

The Industrial Policy Resolution, 1956

The Industrial Policy Resolution, 1956 (IPR, 1956)[11] of the union government defined the industrial policy, which remained in force until the 1991/92 economic liberalization. To achieve the goal of a socialistic pattern of society by means of appropriate economic and social policies, some industries in Schedule A were reserved for the government sector. These industries included atomic energy, heavy electrical plants including hydraulic and steam turbines, coal and lignite mining, mineral oils, and electricity generation and distribution.

However, many foreign-owned companies in the petroleum, manufacturing, and banking sectors were nationalized during the next two decades, and licensing controls were extended to all aspects of industrial development – capacity addition, imports, foreign equity ownership, and sometimes sale price – thereby reducing the competitive spirit of business, the scope for innovation, and the potential to make their operations more efficient.

The public expects cheaper electricity

The mixed signals emanating from this policy were voiced by the chairman of BSES in 1960, "for some time no electric supply company has had the confidence to enter the market for equity capital ... the best way to restore equity values, which alone will attract development capital in the industry, is for the Reasonable Return itself to be upgraded from the present level of the Reserve Bank rate plus 2% to at least the bank rate plus 3%."

While the licensed private suppliers sought marginal improvements in financial returns, in 1960 activists demanded that BSES, which supplied electricity to the suburban areas around Bombay, be municipalized to bring down its electricity charges to the same level as that in the city. More generally, the growing recognition that electricity supply was a quasi 'public good' – an economic input for development – spurred the growth of publicly owned generation, transmission, and distribution capacities. Some of the subsequent growth in electricity supply was also driven by the SEBs charging farmers and domestic users, tariffs often less than even the average cost of supply and looked to the state governments to subsidize the difference between revenue from tariffs and expenditure on supply. To reduce the drain on the state's budget, the SEBs also generated a 'cross subsidy' by charging commercial and industrial users more than the average cost of supply.

Events in faraway UK had been no different. Commenting on the outcomes of the Electricity Supply Act 1926, in Great Britain, R H Coase of the London School of Economics and Politics wrote pithily in 1950,[12] "the electricity supply industry in Great Britain passed into the hands of the State on April 1st, 1948, and private enterprise in that industry was extinguished."

Private electricity supply in India remained frozen till the economic liberalization in 1991/92 deregulated industrial licensing. In the electricity sector this resulted in the re-entry of private enterprise into the space it had vacated earlier. Table 4 illustrates the declining share of 'non-utilities' – a proxy for privately owned captive generation capacity – during 1961 to 1981 and the subsequent revival post 1990–91.

Table 4 Electricity generation shares: utility and non-utility

Plan period	Utility		Non-utility		Total	
	Units (**TWh**)	%	Units (**TWh**)	%	Units (**TWh**)	%
1960–61	16.9	84	3.2	16	20.1	100
1965–66	32.9	90	3.6	10	36.5	100
1970–71	55.8	91	5.4	9	61.2	100
1975–76	79.2	92	6.7	8	85.9	100
1980–81	110.8	93	8.4	7	119.2	100
1985–86	170.4	93	13	7	183.4	100
1990–91	264.3	91	25.1	9	289.4	100
1994–95	351	92	32.1	8	383.1	100

SOURCE CMIE India's Energy Sector, September 1996.

The fiscal cost of public electricity supply

The shift to publicly owned supply was at a significant fiscal cost. The fiscal situation of the SEBs worsened over time. After three decades of their existence, 16 out of 19 SEBs were unable to recover even the cost of supply through tariffs. The average rate of return for the SEBs, which was −9.4% in 1985/86, slid to −13.5% in 1995/96 (Table 5).

Table 5   Annual rates of return in select SEBs (without government subsidy)

State	1985/86 (%)	1990/91 (%)	1995/96 (%)
Andhra Pradesh	6.2	−1	−11.3
Assam	−50	−36.1	−17.2
Bihar	−15.4	−26	−19.1
Delhi	0	0	−16.5
Gujarat	−7.4	−23.6	−20.3
Haryana	−11.7	−14.9	−21.9
Himachal Pradesh	−29.8	−8.6	−1.3
Jammu & Kashmir	0	0	−45.7
Karnataka	−10.3	−2	−2.3
Kerala	3.3	−14.4	−4.8
Madhya Pradesh	−2.1	−11.9	−13
Maharashtra	−11.6	−15.1	−18.6
Meghalaya	−7.7	−34.2	−6.2
Orissa	1.2	3	4.3
Punjab	−12.8	−22.6	−29.3
Rajasthan	−8.1	−11.1	−13.5
Tamil Nadu	−11.8	−15.1	−18.6
Uttar Pradesh	−11.7	−16.2	−12.4
West Bengal	−17.9	−42.8	−21.8
All SEBs	−9.4	−13.9	−13.5

SOURCE ARSEB&ED, Planning Commission in N Govinda Rao, Kalirajan K P, Shand R. *The Economics of Electricity Supply in India*. Macmillan India Limited. 1998.

The highest negative annual rate of return, calculated under the methodology prescribed in the ES Act, 1948 (which defined the asset base on which a defined return above the Reserve Bank rate was allowed) ranged from −50% in Assam to −11.6% in industrialized Maharashtra and −7.4% in Gujarat. Only Andhra Pradesh, Orissa, and Kerala retained positive rates of return. Sadly, by 1990/91 even Andhra Pradesh and Kerala slid into negative rates of return. This lack of business self-sufficiency, and reliance on subsidies from the state governments continue.

Electricity-intensive development

For over three decades, from 1951 to 1985, the per capita consumption of electricity increased from 18 kWh to 229 kWh – an annual increase of 7.8% whereas the economy grew at less than half of that rate, with GDP (constant terms) increasing only by 3.73% annually. The high intensity of energy in economic growth was due to the following factors.

1. The pattern of industrialization favoured heavy industries such as aluminium, steel, cement, fertilizers, heavy machinery, and petrochemicals, which were intended to delink the economy from dependence on import. Growth came at the cost of high energy intensity.
2. The extension of access to electricity to rural areas, including for agricultural use, meant higher consumption. "Between 1951 and 1976 the number of electrified pump sets increased from 200,000 to 3.2 million—a 20% annual growth rate. The census data for 1960 and 1970 show that the proportion of rural establishments using electricity increased from 1.2 to 5.4 per cent over this period, compared with an increase from 14.1 to 18.9 per cent among urban establishments."
3. Progressive electrification of villages increased energy consumption. As against 3061 electrified villages in 1950, electricity had reached 370,322 villages by 1985, representing about two-thirds of the total number of villages.

4. Low generation capacity utilization (a thermal plant load factor of just 48% or less than half of the rated capacity) and low operational efficiency in distribution with line losses (the difference between electricity injected into the grid and the volume billed to customers) amounting to an estimated 19% (1994/95), resulted in energy shortages estimated at 9.3% in 1982/83, 10.8% in 1983/84 and 6.1% in 1984/85 and increased the cost of supply.

Changes in consumption pattern

The pattern of electricity consumption also changed over time. In 1950, 5.5% of the supply was used for power traction (tramways in cities) and only 2% for agriculture. By 1985 the share of traction had dwindled to 2% and that of agriculture increased to 17%, where it remains today.

The share of domestic use increased from 9% to 12% whereas that of commercial use remained constant at about 6%. The share of industry decreased significantly, from 72% in 1950 to 49% in 1985. Some of this decrease is explained by the industry resorting to private captive generation, which increased from 1468 GWh in 1950 to 12,346 GWh in 1985.

Grid supply conditions and quality deteriorated because of underinvestment in transmission and distribution infrastructure. Increasing generation capacity had become a performance benchmark to overcome supply shortages. Capital allocations for improving transmission and distribution capacity suffered in comparison.

The state electricity boards became increasingly fiscally stressed within 25 years of the Electricity (Supply) Act, 1948 coming into force, which had made them the primary conduits for implementing the national power policy. A policy tweak was necessary. Box 1 provides details of publicly owned companies that were incorporated by the union government to add capacity in thermal and hydro generation; finance the planned extension

Box 1 Corporate initiatives in electricity supply of the Government of India

1948 The union government along with the governments of West Bengal and Bihar (now Jharkhand), set up the Damodar Valley Corporation (DVC), a multisector project to manage perennial floods in the Damodar valley, to harness river water for generation of hydroelectric power, and to pursue development of the region.

1951 The Bhakra Beas Management Board, chaired by the union government, with Punjab, Haryana, Himachal Pradesh, and Rajasthan as members, owns and operates the Bhakra (constructed during 1951–63) and Beas (constructed in the 1960s) projects, India's largest interlinked water storage dams, supply 34.5 billion cubic metres of water to irrigate 5.5 million hectares of land and drinking water to Delhi; generated 11.5 billion units of electricity in 2020/21 and assist POSOCO (Power System Operation Corporation Ltd), the national system operator, by ramping up or down from 2500 MW to 400 MW to stabilize grid frequency and provide 'black start' support.

1956 The Neyveli Lignite Corporation was established as a state-government venture to mine lignite in Tamil Nadu. The corporation was transferred to the Government of India with the understanding that Tamil Nadu would be the sole beneficiary of the power generated. The first 50 MW unit of a 600 MW thermal power plant was commissioned in 1962.

1967 Uranium Corporation of India Limited was incorporated under the DAE (Department of Atomic

	Energy) to mine and process uranium for pressurised heavy water reactors.
1967	Electronics Corporation of India Limited was set up under the DAE to develop the communication systems required for India's nuclear programme.
1969	The Rural Electrification Corporation was registered as a public sector financial institution to finance rural electrification schemes and to promote rural electric cooperatives throughout India.
1975	The National Thermal Power Corporation Limited, a fully owned government company, was incorporated to set up large-scale thermal power generation projects. Its first project was in Singrauli in Uttar Pradesh.
1975	The National Hydroelectric Power Corporation Ltd was established to implement large hydro projects. Its first project was at Bairasiul in Himachal Pradesh. Both NTPC and NHPC built long transmission lines connecting existing state grids to evacuate the power they generated.
1976	The North Eastern Electric Power Corporation Ltd was established to plan, develop, and operate hydro power stations in the north-eastern states of India. Its first project was a 200 MW hydro power station in Kopili in Assam.
1986	Power Finance Corporation, a union-government-owned non-banking financial company, was constituted as the nodal agency for developing integrated power development schemes, ultra-mega power projects, and independent transmission projects. Currently the largest such company in net worth with a market share of 20%.

1987	The Nuclear Power Corporation of India Ltd was incorporated under the DAE to design, construct, commission, and operate nuclear power plants. It currently generates 6.8 GW capacity across 22 power generating reactors.
1987	Indian Renewable Energy Development Agency Ltd was established under the Ministry of New and Renewable Energy as a non-banking financial institution for promoting, developing, and financing new and renewable energy and energy-efficiency and energy-conservation projects.
1988	Tehri Hydro Power Development Corporation was constituted as a joint venture between the union government and the Government of Uttar Pradesh and generates about 5 GW comprising hydro, thermal, and renewable energy. In 2020 the union government equity was transferred to the National Thermal Power Corporation.
1989	National Power Transmission Corporation Ltd. (renamed Power Grid Corporation of India Ltd in 1992) was established to develop an integrated national power transmission system network.
2003	Bharatiya Nabhikiya Vidyut Nigam Ltd, a company wholly owned by the DAE was established. The corporation is building a 500 MW fast breeder reactor at Kalpakkam, Tamil Nadu.

of grid power to rural areas; support regional development in the North East, and to promote renewable electricity.

The results were as expected. The resource flow increased, and electricity generation and transmission capacities were increased. Over the next 35 years, SEBS would increasingly focus

on state-level transmission and distribution while the union-government-owned companies became dominant in centralised generation and interstate transmission and grid management.
In the 1990s, even this policy tweak ran out of steam. The union government was forced to reinvite private generation utilities—a business that had been 'extinguished' four decades earlier by law.

The return of private investment: 1985–2020

By the mid-1980s global trends in economic governance had shifted to reducing the control of governments and allowing more room for markets to function in the belief that the efficiency of markets in allocating and managing capital could not be matched by governments. The collapse of Soviet Union in 1989 provided incontrovertible evidence of the downsides of big government.

Governments recognised that spending – particularly if it was based on borrowing, as in India, rather than from a surplus of tax revenues over expenditures, needed to be handled with restraint. In India, particularly in electricity supply, the need for private investment to support government outlays was keenly felt.

The need for private investment in electricity sparked the interest in alternative modes for effective regulation of public private partnerships in infrastructure. By the end of the 1980s the environmental impact of generating electricity from fossil fuels was gaining prominence though the subsequent demonising of coal [see *Coal: phase down or phase out* in this series] would come only three decades later. The role of government in hand holding new renewable electricity like wind power was being recognised as were initiatives to incentivise energy efficiency in electricity generation and supply. In short India was on the threshold of acquiring global scale in electricity supply and increased association of the private sector was one way of getting there.

International comparison of electricity supply

By 1985, India had installed 42.6 GW of electricity capacity and generated 186 TWh of electricity. Table 6 shows how this achievement in electricity generation compares with some other countries over the period 1985 and 2020.

Table **6** Electricity generation: global rank by volume

1985			2020			1985–2020
Rank	Country	Generation (**TWh**)	Rank	Country	Generation (**TWh**)	Generation increase (**% per year**)@
1	US	2657	1	China	7799	8.8
2	Russia	962	2	US	4287	1.4
3	Japan	672	3	India	1561	6.3
4	Germany	523	4	Russia	1085	0.4
5	Canada	459	5	Japan	1005	1.2
6	China	411	6	Canada	644	1.0
7	France	344	7	Brazil	620	3.3
8	UK	298	8	S. Korea	574	6.5
9	Ukraine	272	9	Germany	572	0.3
10	Brazil	194	10	France	525	1.2
11	India	186	11	Saudi Arabia	341	5.5
22	S. Korea	63				
28	Saudi Arabia	52				

SOURCE BP Energy Statistics 2021

@   Increase in generation refers to annualised increase by 2020 relative to the level in 1985

In 1985 India was the 11th largest generator of electricity. By 2020 it had become the 3rd largest generator, after China and the United

States. Brazil improved its rank from 10th in 1985 to 7th in 2020; South Korea, from 22nd to 8th; and Saudi Arabia, from 28th to 11th.

To be sure, this simplistic ranking masks far more important metrics, namely the per capita supply of electricity, which measures its average availability for citizens and its cost and quality. Generation per person is a better metric of development in which India continues to lag other advanced economies and China (Table 7). Generation per person in India in 2020 at 1131 kWh was 4.7 times higher than generation per capita in 1985 of 237 kWh. However, it was just one-third of the global average annual generation per person of 3464 kWh.

Table 7 Generation per person (kWh/yr)

Country	1985	2020
US	11,048	12,952
France	6266	8040
Germany	6731	6826
UK	5284	4610
China	382	5418
India	237	1131
World average	2028	3464

SOURCE (a) Electricity Supply – BP energy statistics.
(b) Population – www.populationpyramid.net/world

The United States continues to generate the most electricity per person, with France and Germany not far behind. In the UK generation per person decreased by about 10% between 1985 and 2020 and it was overtaken by China, which now exceeds European per-person generation levels. India lags significantly in per capita availability. It also has the lowest per capita income in this select list of economies. In the not-so-distant future, the share of green or renewable electricity in the total energy mix will be the most important metric for development as controlling carbon emissions gains urgency.

Profligate use of electricity shunned

During 1985–2020 the perception that electricity intensity is the driver behind economic development changed; instead, energy conservation and energy efficiency became the instruments to reduce the emissions of carbon dioxide and other GHGs associated with the use of fossil fuels and thereby limit the increase in global mean temperature to less than 2 °C.

Since the 1992 Kyoto Protocol, which set targets to limit emissions of GHGs in advanced economies, consumption patterns of electricity have changed. Such binding constraints did not apply to China or India; in fact, China increased its fossil fuel dominated generation by 9% annually during this period.

India too increased its generation capacity by 6.3% annually, the second highest growth rate in generation after China and South Korea (Table 6). Unlike during 1950–1985, when GDP growth in India had trailed energy consumption, real GDP during 1985–2020 grew by 6.2% annually, thereby reversing the energy intensity of GDP growth compared to the first 35 years from 1950. One reason is the rapid growth of the low-energy-intensity services sector, which accounted for 49% of GDP in 2020, lowering the share of the energy-intensive manufacturing sector to 16%, much lower than in the past.

Incentives for electricity reform

The incentive for electricity reforms in India was twofold.

First, the parlous state of SEBs presented a stark picture of the general inefficiency of the tightly regulated and controlled Indian economy in 1985. The growing burden of power subsidies put pressure on the union and the state government finances and reduced the public investments available for enhancing the capacity and extending the reach of the supply network.

Second, the experience of the UK in the immediate past, as in 1948, strongly influenced the course of events in India. The Conservative party led by Margaret Thatcher legislated the UK Electricity Act 1989, for reforming and restructuring the electricity industry as a continuation of its broader agenda of liberalisation. This model of electricity reform introduced competition in supply by unbundling of electricity supply chains into generation, transmission, distribution, and retail supply and privatizing state-owned enterprises. It constituted independent regulatory agencies to apply market principles to determine the cost of supply and cost reflective tariffs to stabilise the industry financially. Open access in retail and determination of the market clearing price for bulk supply in a power pool with generators and bulk buyers bidding for power was the goal.[13]

Union government's initiatives for structural reform in electricity

Continuing with the policy of direct intervention in the generation of electricity, the union government next turned its attention to transmission and independent system operation.

Institutional reforms in transmission

The report of the Rajadhyaksha committee on power sector reforms, submitted in 1980, recommended the need for a national grid – a resilient, high-voltage,[14] national electricity transmission network – which could integrate the splintered state and regional grids for seamless transfer of surplus electricity and reduction in the cost of supply by transmitting electricity directly from pit-head, coal-based generating plants to consuming centres rather than transporting coal to distant load-based generators. A decade later, in 1989, the National Power Transmission Corporation Ltd was established to plan, execute, operate, and maintain such a high voltage transmission system. It was renamed the Power Grid Corporation of India Ltd (PGCIL) in 1992.

Between 1991 and 1993, the transmission assets owned by the union government – the assets of such entities as the Nuclear Power Corporation of India Ltd, Neyveli Lignite Corporation, NTPC, NHPC, NEEPCO, and the Tehri Hydro Development Corporation – were all transferred to PGCIL.

In 1994 PGCIL was mandated to manage the national load dispatch. In 1996 the five regional load dispatch centres were formally transferred to PGCIL, which was notified as the Central Transmission Utility (CTU) by the Government of India under the Electricity Regulatory Commissions Act, 1998. The provisions of this legislation were subsumed into the Electricity Act, 2003. In 2009 the CTU established the National Load Dispatch Centre (NLDC) as a precursor to creating a separate legal entity to manage system operations – scheduling of dispatch, monitoring grid conditions, managing frequency and securing grid stability. NLDC was hived off to POSOCO, a subsidiary of Powergrid/PGCIL, thereby creating a desirable regulatory wall between the transmission business and the entity responsibility for grid management.

Powergrid has grown from strength to strength. In 2013 it connected India and Bangladesh through a 500 MW transmission link enabling cross-border electricity transfers. In the same year it fully synchronized the national grid by completing the Raichur–Solapur transmission link. In 2016 it commissioned a Nepal–India transmission link.

Powergrid, as a central transmission utility, has actively enhanced competition in transmission by endorsing the entry of private transmission entities in the national grid, who have a share of 7% in terms of length in transmission lines and of 4% in terms of the number of substations managed by them. The Electricity Act, 2003, requires such endorsement by the central transmission utility before private investment in transmission is licensed by CERC (Central Electricity Regulatory Commission).

Open sesame: private power policy 1991/92

By the end of the 1990s, the state power entities owned by the union government were primed to function within a reformed power sector architecture. The only problem was that of insufficient funds for the expansion of electricity assets to meet the highly aspirational targets. One option was to use the 1991/92 reform momentum unleashed by the economic reforms to also push for reforming the electricity sector. Opening generation to private investment seemed an obvious solution to energy shortages that have always dogged India.

The generation of electricity was thrown open to private investors through the Statement of Industrial Policy, 1991,[15] overruling the prohibition imposed in the Industrial Policy Resolution, 1956. Simultaneously the FDI (foreign direct investment) regime was liberalized and the prohibition on majority ownership by a foreign entity was removed.

Ending the policy of protecting domestic investors from foreign competition signalled that India had shed the inhibitions of the post-Independence era when the government had barred foreign investment and nationalized several foreign-owned companies thinking that it could not regulate these entities otherwise. The expectation from liberalization was that, as in Russia and China, foreign investors would rush to invest in the infrastructure sector. These expectations were, however, belied. Most foreign investors found the regulatory ecosystem unconducive for profitable business in the power generation sector.

Three main near-unsurmountable barriers stood in the way of such investment. First, no SEB had a stand-alone, viable revenue profile. Investors were therefore compelled to enter into long-term sale agreements for power with specific SEBs backed by the state governments. Second, fuel had to be arranged by the investor— no easy task in an economy in which all the fuel suppliers were public-sector entities directly controlled by the union government. Third, the SEBs remained monopoly buyers of bulk power in their respective areas, which meant that private investors could not go around the SEBs to specific large-scale consumers or selectively target such richer cities as Mumbai, Delhi, Bangalore, or Chennai.

The need for electricity supply markets

A competitive market for electricity was needed. It could be created either by privatising the existing state-owned entities or the barriers to their efficient functioning could be incrementally removed.

Privatizing the electricity industry was never an option. Distribution was owned by the state governments, and the complexities of taking on the collective might of the industry unions, supported by over a million employees, required ideological commitment – as in the UK – which was absent. The term 'socialist' had been added to the preamble of the Constitution in 1976 through a constitutional amendment to better reflect how India had developed since independence.

Not surprisingly, the preferred political solution in the 1990s was one of incremental reform. 'Reform by stealth' acquired a connotation of smart reform in the generally unsupportive political climate. The 'liberalization package' amended the Electricity (Supply) Act, 1948, allowing private generators to quietly infiltrate the citadel of state-owned public utilities. Simultaneously, an enabling ecosystem was encouraged via multilateral support.

Mega power policy, 1995

Recognizing the financial constraints in distribution and the need to continue to subsidize select customer segments such as agriculture and small domestic users, the state provided incentives to private generators to encourage capacity development. The Mega Power Policy, 1995, targeted thermal plants with capacities above 1000 MW (above 750 MW in a few special states) and hydroelectricity plants of capacity above 500 MW (above 350 MW in a few special states), which supplied electricity to more than one state and were therefore regulated by the union government.

The most important incentive was zero duty for import of equipment, eligibility for deemed export benefits, and a price preference of 15% for domestic bidders, on the condition that the states procuring the power must have constituted regulatory commissions and must agree in principle to privatize distribution in all the million-plus cities.[16] This condition was relaxed in 2009[17] – the power-purchasing states had only to agree in principle to distribution reforms. The applicability of the incentives was broadened to include even intra-state supply project and brown-field expansions.

Ultra-mega power projects policy

The ultra-mega power projects policy was launched in 2005 targeting projects with capacities of 2000 MW or higher using super critical technology. The Power Finance Corporation (PFC), a public-sector financing company under the Ministry of Power, was designated as the lead agency. The PFC would form SPVs (special purpose vehicles) to get all the requisite permissions in advance, prior to the bidding, and transfer a running SPV to the winning bidder, providing a painless 'turnkey' entry to private investors.

Out of nine projects initially shortlisted, four were selected but only two were finally commissioned – Sasan in Madhya Pradesh, which used domestic coal, was won by Reliance Power Ltd at a levelized cost of Rs 1.196/kWh, and Mundra in Gujarat, which used imported coal, was won by Tata Power at a levelized cost of Rs 2.261/kWh. Of the two other projects, the winning bidder in 2008 for the Krishnapatnam UMPP stopped work citing cost escalation in imported Indonesian coal on which the project was based. As for Tilaiya, the UMPP project in Jharkhand, also won by the same company in 2009, a notice terminating the power purchase agreement was submitted on the grounds that the required land had not been transferred to the company in time.

Table 8 provides the generation capacity additions by union, state, and private entities. The private sector's response to the successive measures offering fiscal and procedural support, was impressive. Its share in capacity additions in the private sector increased to comprise 27% of capacity additions during 2007–12 and 55% of capacity addition during 2013–17 before decreasing subsequently.

Table **8** Generation capacity addition (**GW**): 2007 to 2020

Source	2007–12	2013–2017	2018	2019	2020
Centre	60	21	4	2	4
State	86	25	2	2	3
Private	54	54	4	2	0
Total	200	99	10	6	7
Renewable energy	25	49	12	9	9

SOURCE 2007–12 – 12th plan document; 2013 onwards – MoP; Renewable energy capacity addition – MNRE

The slowdown of private investment in conventional power capacity can be ascribed to three reasons.

First, economic growth slowed from 2017/18, which lowered the incremental demand for electricity leading to a period when India had surplus power even before the COVID-19 pandemic struck in 2020/21.

Second, in 2015, the union government launched an ambitious programme to support solar energy generation, which diverted the interest of private investors to renewable energy bolstered by the global sentiment in favour of green power after the Paris climate meet in that year. Table 8 illustrates how capacity addition in renewable electricity increased from an average of 5 GW per year during 2007–12 to nearly 10 GW per year during 2013–17, further increasing to a peak of 12 GW in 2018 before slowing to

9 GW in 2019 and 2020. By March 2022, privately owned renewable generation capacity accounted for 107 GW, or 95% of the total renewable energy capacity of 111.5 GW.

Third, the continuing financial instability in the discoms soured investment sentiment, a disenchantment that also negatively impacted investment in power generation because of increasing volumes of unpaid bills for the power supplied by them to distribution companies. Arrears to generators owed by discoms were assessed at Rs 1.08 trillion in May 2022, of which 57% were payable to private generators producing electricity the conventional way (using fossil fuels), 20% to private renewable electricity generators and 23% to generators in the public sector (agencies of the union or state governments).

Distribution reform

The union government's entry into distribution reform was indirect, possibly aligned with the concurrent constitutional mandate that gives the state governments pride of place in local retail supply whereas interstate supply is the union government's mandate. The union government also supported state-level reforms, tacitly, through support from multilateral and bilateral donors such as the World Bank, the Asian Development Bank, and Britain's Department for International Development (DfID, now the Foreign, Commonwealth & Development Office). Consequently, even before the union government could formalize the process of distribution reforms, three states – Orissa (now Odisha) (1995), Haryana (1997), and Andhra Pradesh (1998), all with charismatic chief ministers – decided to take the lead and transform their electricity sectors following the neo-liberal reforms template which had originated in the UK.

All the three states enacted legislations creating autonomous regulators to determine the tariffs for generation, transmission, and distribution within the state and to regulate the standards of supply. The assets related to generation, transmission and distribution were unbundled from the SEBs and smaller distribution companies were set up; in Odisha these were privatized unsuccessfully. A franchisee model also failed. By 2022 all four discoms had again been privatised with Tata Power as the majority owner (Box 2). In Andhra Pradesh discoms were to be privatized but the efforts suffered a setback when the 'reformist' government of Chandrababu Naidu was defeated in the 2004 state elections; Haryana never committed to privatizing distribution in the first place. Of four other states which had also subsequently passed legislations for reform – Karnataka and Uttar Pradesh in 1999 and Delhi and Madhya Pradesh in 2000 – only Delhi went beyond forming an SERC (state electricity regulatory commission) and unbundling to privatization, which, in sharp contrast to Odisha, has been an unqualified success.

> Box 2 Odisha: early start at reform stumbles
>
> Odisha's charismatic chief minister Biju Patnaik, a war hero, a pilot, and an Odisha grandee, shepherded the reform. The political motive was that better electricity supply could fast-track industrialization, particularly around the considerable mineral wealth of Odisha. Multilateral agencies – the World Bank in this case – took it to be winner. A very low agricultural load of 6%, cheap hydropower, and a high industrial load promised adequate cross subsidy. Better management could fix the problem of high aggregate technical and commercial loss of 60%. By 1995 the reform legislation was in place, the Orissa Electricity Regulatory Commission was constituted, and the SEB was unbundled into a publicly owned generation company, GRIDCO, a transmission company, and four distribution companies which were privatized. One, with low industrial load, never

stabilized and AES, the American company, surrendered its license in 2001. The other two companies struggled: BSES, a license holder, sold the license to Reliance in 2003, but Reliance too incurred losses while aggregate technical and commercial losses remained high at 40%. The state regulatory commission revoked Reliance's license in 2013 for non-compliance. The company argued in the Supreme Court in 2015 that the OERC had failed to revise the tariff in line with increasing costs. The entire distribution business consequently reverted to GRIDCO, undoing the unbundling. The only difference was that the Talcher thermal power plant had been sold to the National Thermal Power Corporation to raise funds for GRIDCO early on. The aggregate technical and commercial losses had been halved (29% in 2019/20) from the earlier 60%, and access to electricity had been extended in rural areas. One company was then managed by GRIDCO and the other three managed by private franchisees which improved billing and collection efficiency significantly, but none could afford the investment required to reduce the losses to the targeted 15% from 25%–29% in 2019/20.

SOURCE Das M and Nayak M. 2018. Endless restructuring of the power sector in Odisha: A Sisyphean tale? pp. 193–214 in *Mapping Power*, edited by N K Dubash, S S Kale, and R Bharvikar. New Delhi: Oxford University Press. 400 pp.

Meanwhile the reforms continued. The Electricity Regulatory Commissions Act, 1998, was legislated by the union government.

The Electricity Regulatory Commissions Act, 1998
The Electricity Regulatory Commissions Act, 1998, was cleverly drafted to diffuse any likely opposition from state governments

given that electricity supply is a concurrent subject under the constitution and the states had been dominant partners since 1948. This was about to change.

First, a central regulator (Central Electricity Regulatory Commission) was established under the act to determine the tariff for all central generation and transmission assets and for any cross-border transfer of electricity. Additionally, any generation or transmission activity, such as supplying electricity to more than one state, came under the mandate of the central regulator. The union government acted magnanimously in externalizing its regulatory powers over its electricity assets to the CERC. This created the ground for state governments to follow suit.

Second, the act provided for state electricity regulatory commissions to determine the tariffs for generation, transmission, distribution, and retail supply per the principles mentioned in the act and sub-regulations. These required tariffs to reflect efficient costs. The cost of subsidy was to be borne by the government directing the subsidy payment, and no customer class could be discriminated against. Basic market principles were to govern the price setting.

Third, the act mandated the regulators to work towards creating a competitive and efficient industry in the interest of consumers.

Following the passing of this legislation by the union government, several states committed to reform their electricity sectors incentivised by specially curated reform packages financed by the World Bank, the Asian Development Bank, and associated bilateral donors. Adopting the institutional change of transferring regulation to an autonomous regulator was the easiest step and near-universally complied with. Unbundling the distribution business from transmission and corporatizing the newly created constituents took longer, with states evolving their own models of unbundling.

Delhi's privatised discoms
Privatization of distribution was successful in Delhi. Other than in the cantonment area and the Lutyens Delhi zone, retail supply and distribution is managed by three private discoms since 2002 with continuing and significant improvements in the quality and stability of supply. High average personal incomes of customers, a well-funded state government, and the absence of the burden of heavily subsidized agricultural supply have all contributed to creating and facilitating an efficient, welfare-oriented electricity supply ecosystem. Most states have not even attempted to privatize distribution on a large scale although localized experiments with franchising small areas continue.

The Electricity Act, 2003
The legislative agenda was pursued with increasing vigour. Within five years of the 1998 act, a new legislation was enacted in 2003, integrating the features of the Electricity Act, 1910; the Electricity (Supply) Act, 1948; and the Electricity Regulatory Commissions Act, 1998.

The outcomes of reforms
The end of supply shortages
The continuing energy shortages and peak shortages have ended.[18]

In 2019/20 the energy shortage was only 0.4% and the peak shortage 0.7%. This indicates that either the generation capacity addition has overcome the sluggish growth earlier relative to increase in demand or that demand itself has slowed down significantly.

Table 9 provides a snapshot of the trend of operational efficiencies in the electricity supply industry in India.

Generation capacity utilization lower than optimal
Plant load factor (PLF), a metric of the extent, to which installed capacity is being used for generating electricity, was a satisfactory

75% in 2010/11 but had decreased to 56% by 2019/20 because of additional generation capacity coming online and lower demand following the slowing down of economic growth even before the COVID-19 pandemic.

Table 9   Operational efficiency in electricity supply: 1995 to 2020

Indicator	Unit	1995–96	2000–01	2010–11	2019–20
Energy shortage	%	−9.2	−7.8	−8.5	−0.4
Peak shortage	%	−16.5	−13	−9.8	−0.7
Discoms losses (including subsidy)	Rs billion	−71.3	−202.2	−538.9	−867.0
Subsidy received	Rs billion	16.9	370.4	202.9	1135
Payables to generators/ electrical suppliers	Rs billion	104.6	293.1	686.1	259.1
Plant Load Factor (Coal generation)	%	60	69	75	56
Technical and commercial line losses in Discoms	%	19	23	26	21

SOURCES

1994–95 data: Planning Commission, in N Govinda Rao, Kalirajan K P, Shand R. *The Economics of Electricity Supply in India*. Macmillan India Limited. 1998.
2000/01 data: Annual Report on working of SEBs, 2000–01. Planning Commission
2010/11 and 2019/20: Report on Performance of Power Utilities, 2020–21. PFC.

Transmission shines

Transmission is the brightest segment in the supply chain with efficient expansion and management of the national grid. Grid discipline has improved, including in regional and state grids, since implementation, in 2002, of the availability-based tariff which penalizes generators feeding more electricity into the grid than scheduled in advance and similarly unscheduled withdrawal of electricity by discoms.

Private power exchanges

The establishment of two private power markets for bulk supply, which transact approximately 9% of annual electricity supply and manage trade in renewable energy certificates, acts as a base for the future expansion of 'open access' in retail. In August 2022 a third exchange will become operational. Nepal has joined the power market for buying electricity as well as to sell electricity generated in Nepal.[19]

Credit should be given to the efforts of the union government to get the private sector to invest more in generation, first through the mega power policy and then through the ultra-mega power policy of 2012. By 2022 the share of private generation has increased to 36% of the total conventional generation capacity and 96% of the renewable energy capacity.

Discoms: the weakest link in the value chain

Distribution and retail supply remain a weak link in the chain—a sink that soaks up government subsidies and fiscal support although the extent varies greatly across states. The subsidy provided by state governments to the tune of Rs 867 billion in 2019/20 equalled 2.8% of the total state revenue expenditure for that year of Rs 30.7 trillion and 0.4% of GDP (Rs 203.5 trillion at current prices).

Retail tariff fails to reflect a reasonable cost of supply

Despite regular payments of subsidy – excluding such one-time programme grants as debt relief or other incentives – the aggregate revenue at Rs 6.16/kWh still meets only about 90% of the average cost with a gap of Rs 0.60 for every unit of electricity supplied. This reveals the continuing financial imbalance in the distribution segment, despite three attempts by the union government, in 2001, 2012 and 2015, to incentivize reduction in ATC (aggregate technical and commercial) losses by utilities and fiscal restructuring to reduce the burden of paying the interest on their accumulated debt. State governments were encouraged to acquire the outstanding debt of publicly owned discoms to reduce their financial liabilities. To finance this debt acquisition the union government raised the debt ceiling of state governments, monitored by the Reserve Bank of India. This enabled states to acquire the discom debt by borrowing from the markets at lower cost than the interest payable on the discom debt, whilst also lowering the amount spent on future interest outgo and possibly extended the repayment period.

Financial stability in discoms impacts upstream suppliers
Financial stress in discoms leads to mounting dues to power generators, transmission service providers, and coal suppliers – all majorly owned by the union government – which, in turn, fund the delayed payments by increasing their working capital requirements, the ultimate result being costlier supply.

High -aggregate technical and commercial losses
The aggregate technical and commercial losses in discoms, a blunt but useful measure of operational efficiency, remain at 21% (2019/20), well above the prescribed norm of 15%. The accumulated net loss to state-owned discoms in 2019/20 was Rs 5.1 trillion, equal to their accumulated debt of Rs 5.14 trillion.

The unfinished reforms agenda
Reforms during 1985–2022 were principally focused on unbundling the electricity supply industry, introducing autonomous regulation at the union and state levels, corporatizing the unbundled entities to standardize the best-fit managerial and accounting practices, and incentivize private investment in generation.

The agenda of making distribution and retail supply productive and profitable remains incomplete. The inefficient and uneconomic overcharging of industrial, commercial, and large-domestic consumers to cross subsidize agriculture and small-domestic consumers also continues. Subsidy, both direct and implicit (state-guaranteed debt to cover commercial losses) accounts for about 10% of the cost of supply (5% direct subsidy and 5% implicit subsidy).[20]

This burden of subsidy is unsustainable. It will continue to discourage private investment and delay the transition from long-term power purchase agreements to market-based purchase of power from power exchanges by discoms. Financial instability in discoms also implies continued state control over the sector, negating the potential for efficiency enhancement from competition, retail choice and transition to a regime of light-handed regulation.

Planned electricity development: supply-side triumph or copycat industrial policy?

It has been a momentous journey for electricity in India. In 1950, when planning started, the target was to provide electricity to all. At that time, the total consumption was 5.6 TWh with barely 0.7 million households electrified. By 2020 consumption increased to 1248 TWh powering a nearly US$3 trillion diversified economy producing commodities, manufacturing primary, intermediate and final products, offering financial and digital services, supplying electricity to more than 200 million homes and farms and a nearly 100% electrified railway service. More important, to do so at low per-capita incomes is a significant achievement.

Putting industrial development above basic social and human needs

How appropriate was it for India, in the 1950s, to adopt capital intensive supply-side interventions from the government's budget under the planned economy framework?

The state electricity boards were expected to be efficient and to coordinate electricity supply in the respective states. But they failed to overcome the shortfall in supply, which became worrisome by the early 1970s. The planned response was simply to shift the burden of borrowing and capital investment from the states to the union government. New public sector companies were set up to increase generation capacity without bridging the gap between low retail tariffs and the high cost of supplied electricity. Even this strategy faced obstacles within two decades. Finally in the late 1990s the government was compelled to resort to seeking private investment in generation to outsource the task of lumpy capital investments to more competitive private owners. Currently (in 2022) one half of the generation capacity, including renewable capacity, is privately owned.

Throwing out the baby with the bathwater

Would things have been different if we had simply built on the institutional framework that existed in 1947, namely the Indian Electricity Act, 1910? This act permitted private investment in electricity supply. All that was needed was to subsidize the lumpy investments, which private investors would have been wary of making because of uncertain demand. Templates for such cooperation already existed.

The earliest template for the PPP (public–private partnership) mode was the TISCO (Tata Iron and Steel Company) plant in Jamshedpur, which was established to provide steel rails to meet the growing demands of Indian Railways. In fact, the commissioning of a state-built link line to the private-sector plant was crucial to evacuate supply from the plant. In 1908 a conscious decision was made to retain control, within India, of the new enterprise, by seeking out Indian investors. Despite warnings that India could not afford a flotation of this size, some 8000 Indian investors came forward and the whole share issue was taken up (TISCO history) — private Indian investors were both willing and able to invest if the returns were attractive.

Nearly half a century later, over the period of the first five-year plan (1951–56), Rs 2.6 billion of public funds was committed for the development of 1 GW of power generation capacity, adding more than 50% to the 1.7 GW inherited from the colonial period. This public outlay could have been substantially reduced had a PPP-type financing been attempted with government funds used merely to lower the risks instead of putting the entire financial burden on the budget. Unlike the telegraph or irrigation, the users of which were either the state or the farmers, electricity was already being consumed by industries, shops, and homes, all of which would have provided a ready market for reliable supply. The state may have believed that profits from commercial electricity supply could be internalized and used for providing cheap electricity to other customers such as farmers or to the poor.

One cannot blame the state for the public sector orientation of its plans in the early days following independence (1950s to 1960s) during which massive projects combining irrigation and power supply were implemented, which paved the way for food security attained through the Green Revolution. But it was a costly mistake from the mid-1970s onwards to believe that the business of government could ever be to do business.

Had the nationalization of industries and banking in the mid-1970s never happened, the 'lost years' in economic development until the 1991 liberalization could instead have been a period of rapid growth, giving us a head start over China, instead of an embedded two-decade lag in economic growth as today.

Government should have adhered to the maxim that the business of government is to raise revenues for development through fair and progressive tax levied on citizens and corporate entities and prudent borrowing and to use such revenues for all developmental needs, including paying for supply of power to the poor or others in remote areas. The task of providing the services could and should have been left to the private sector: facilitating the supply of electricity should never have extended to financing and managing its production.

Nationalization of private industry and utilities had more to do with ideology and little to do with fears of the inadequacy of Indian entrepreneurs to finance or provide utility services. This is best illustrated by the case of Air India founded by the Tata group. It was a successful aviation business when it was nationalized in 1953. It subsisted on government doles for the next seven decades until 2022, when it was re-privatized and was bought back by the Tata group.

Misallocation of public funds

That capital was misallocated is evident from the fact that despite a doubling of saving and investment rates, growth was slow and even declined during 1950–1980. Low productivity of investment was the proximate reason for increasing investment output ratios and low growth.[21] Table 10 illustrates the poor outcomes of public investment in industries.

Table **10** Misguided investment strategy

Period	Growth rate % Per year	Investment % GDP	Capital/output %
1952–1960	3.5 (1.7)	11.5	3.2
1961–1970	4.0 (1.8)	14.6	3.7
1971–1980	3.0 (0.7)	17.5	5.8

NOTE Figures in brackets – per capita growth rate
SOURCE Joshi V. 2017. *India's Long Road: the search for prosperity*. Oxford: Oxford University Press. 347 pp.

Using life expectancy as a proxy for human development also presents a stark picture of neglect compared to China, Vietnam, or the world average. China improved on low life expectancy sharply in the 1960s whilst India remained well below even the world average (Figure 1).

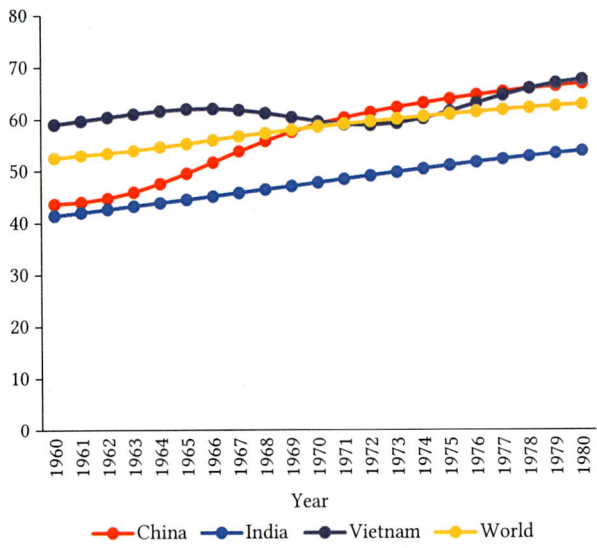

Figure **1** International comparison of life expectancy (average years)

Even industrial manufacturing failed to benefit from the public financing model of planning led by the public sector. Industrial growth from 1952 to the mid-1960s did respond to the stimulus provided by investment in big planned public-sector projects. However, from the late 1960s to the early 1980s, industrial growth decelerated to 4.8% annually. Investment in the public sector and severe constraints to licensing private manufacturing did not provide the disaggregated stimulus needed for Indian manufacturing to grow into a competitive force. This was despite the planned attempt to make India, a low-income economy until recently, 'resemble' middle- and high-income economies, which enjoy a diversified industrial base and competitive edge.

In 1984, India was ranked 43rd, out of 71 economies, in industrial growth. Significantly, the faster growing economies spanned a diverse group, spread across a wide spectrum of geographical size, socio-political system, or degree of openness to the world economy. Economies with a comparable industrial base and structure that grew significantly were China, Brazil, Yugoslavia, Mexico, and Turkey, and 'star performers' such as Taiwan, Korea, Singapore, and Hong Kong were even further ahead.[22]

Borrowed templates

Admittedly, to defend the state's thinking, direct state intervention in electricity and coal was the norm at the time even in the UK, the economy with which India was most familiar. More important, Soviet Union provided a recent and muscular example of state might, much as China does today, of state control and public investment. These experiences would have looked attractive to a fledgling Indian government, eager to provide all the public services that the colonial government had denied to Indians.

Soviet assistance during this period encouraged 'copycat' development of the type they had pursued, which also allowed the Soviets to export equipment and heavy machinery as aid to India. Developing and strengthening the public sector was also an ideological link anchoring India to a potentially communist future, at least until Soviet Union itself became disillusioned with it.[23] The prolonged pessimism about the private sector and early reliance on state controls and investment had a long-term deleterious impact on human development, productivity, and growth. Notably, immediately after independence, there was an attempt to end private ownership of land—the communists preferred collectivism whereas the social reformers liked co-operatives as more socially responsible models for land holding.

Luckily, better sense prevailed. To align with the political rhetoric the easily evaded land ceilings were prescribed in subsequent legislation for land redistribution to the landless, a politically astute but otherwise questionable objective for a land-scarce country with land fragmentation, not concentration, being a problem.[24] The result is that agriculture has remained a private mode of production, unlike in Russia and China. Land-based occupations – agriculture, contract work, and rural services – provide sustenance to more than half of India's 270 million households well beyond the share of agriculture in GDP, which is about 15%. The success of the Green Revolution of the 1960s and 1970s was premised on the entrepreneurship of private farmers who adopted new crop varieties and agricultural production methods to secure the country against famines, which had plagued India until the mid-1960s.

If nothing else, the success of private farming in India with the government providing R&D and input support should have alerted government planners that the industrial and utility sector could witness a similar transformation through private management and investment.

Path dependency

Continuation of the planned public-sector-driven model of electricity development can only be explained in terms of comfortable path dependencies, which militate against transformative change because of the resulting opposition from those who fear loss of embedded privileges. Pragmatism in abandoning the socialist ideological imperative for state ownership of productive assets while retaining the levers of state control over the sector has not served the sector or its customers well.

The fact that half of the generation and most of the distribution entities are still owned by the public sector speaks for itself. The focus on competition in the Electricity Act, 2003, is not aligned with the heavy dominance of the public sector in the electricity industry. The fact that privatization is not considered necessary for competitive utilities illustrates the shallow foundations for competition in the sector.

We seem to have shaken off our ideological imprisonment of the post-colonial period to unquestioningly embrace industrial templates emerging from advanced economies. But we have developed a dependence on the paths we have charted ourselves, since independence, of soft public-sector budgets, a willingness to sacrifice techno-economic logic for options dictated by short-term political economy expediency centred on political survival of elected governments and dilution of public interest to safeguard the interests of the ruling party.

Economic growth and competitiveness

None of this means that the electricity supply industry will be a constraint to the future growth of India. However, it does mean that we will remain non-competitive in either the cost or the quality of electricity delivered unless reforms to unplug the embedded inefficiencies are fast-forwarded. Consider, for instance, the implementation of competition and choice in retail supply, a measure that can enhance industrial competitiveness. This reform will remain hostage to the loss suffered by the public-sector discoms in supplying power at low tariffs to farmers and smaller households.

States with higher per-capita incomes are likely to face less resistance to routine increases in electricity tariffs. Such states also benefit from higher tax receipts, which enlarge their fiscal capacity to usher in welfare measures. As customers become richer, electrical appliances tend to become part of their lifestyles—refrigerators, air conditioners, electric vehicles, computers, gaming consoles, smart phones, and microwave and electric ovens. Customers with high electricity demand prioritize reliability and quality of grid supply over tariffs because the cost of back-up in the form of locally generated power is at least double the grid tariff. The tax earned by the state government from the sale of electronic devices in high-per-capita income states enables the state to subsidize electricity supply to the most vulnerable households.

A good example is Delhi. In 2020 the ruling Aam Admi Party (AAP) made electricity consumption up to 200 kWh a month free of cost and offered a rebate of Rs 800 on the next 200 kWh/month. Delhi has minimal agricultural load, but a sizable population lives in informal habitats, where providing free electricity, as a liberal version of the equity enhancing 'lifeline' concept of public goods supply, instantly builds political support.

With buoyant revenues from a population with the highest average per-capita income in the country, the Delhi government could afford this welfare measure, which was estimated to cost Rs 32 billion, or 4.2% of its Rs 758 billion annual budget.

New pathologies

The AAP now also rules Punjab. It has promised to implement a similar scheme of free electricity up to 300 kWh a month, which means even a home using a 1-tonne air conditioner sparingly, a washing machine or a microwave oven occasionally, need not pay for the power supplied to it. The fiscal consequences could be potentially destabilizing, coming as they do pan-caked on to the Rs 100 billion the Punjab government already spends on electricity subsidy, 70% of which benefits farmers.

Also consider the implications for future energy consumption and carbon emissions if air conditioners are made to cost the least upfront to the buyer with no thought given to the energy efficiency of the appliance. It would undo all the good work done by the Bureau of Energy Efficiency on labelling appliances by their energy efficiency. On all these counts it is preferable to depoliticize electricity supply. The first step in doing so is to privatize discoms with cast-iron guarantees of non-interference from the state in commercial issues.

Formal economywide planning has ended but its hand in the public sector remains more robust in India than is desirable from the economic perspective of building an efficient, competitive, and forward-looking electricity sector.

Unresolved issues in electricity supply
Electricity suppliers: too few or too many?

As on 31 March 2020, there were 150 companies, power corporations, and management boards under central, state, or joint partnership in India in generation, transmission, and distribution apart from 11 electricity departments in the states and union territories, 34 power trading companies, and 144 private licensees including electric supply co-operative societies.[25]

Figure 2 shows the growth in the number of licensees due to the reform acts, namely the Electricity Regulatory Commissions Act, 1998 and the Electricity Act, 2003.

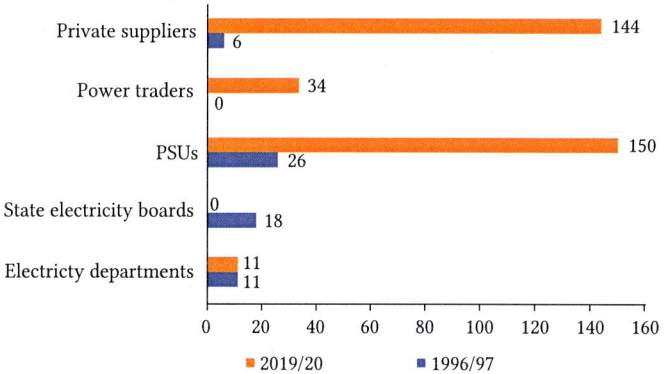

Figure 2 Numbers of different categories of suppliers of electricity: 1996/97 versus 2019/20

Power trading companies are an entirely new class of licensees. They engage in the wholesale business of buying and selling 'blocks' of power through the day on the two power trading exchanges, like brokers selling shares on the stock market.

Unlike shares, though, each 'block' of electricity bought or sold is tagged for delivery at a specific time – at 15-minute intervals. Not delivering the committed supply or not accepting the committed demand invites a dual penalty specified by the power exchange and POSOCO, because both want to ensure that the physical balance of supply and demand of electricity, agreed upon a day ahead, is rigidly maintained.

State electricity boards were created under the Electricity (Supply) Act, 1948. They no longer exist as they were unbundled into separate corporations as part of the reforms, thereby increasing the number of publicly owned corporations from a mere 26 in 1996/97 to 150 in 2019/20.

Private generation entities have transformed the sector, and their number has been growing explosively from just six in 1996/97 to 144 by 2019/20. Many were set up under the mega or ultra-mega power policies along with suppliers of renewable energy, namely wind, solar, and small hydro.

Transmission is dominated by Powergrid, a listed public limited company, majority shares of which are owned by the union government though independent transmission providers are being encouraged, at regional and state levels, the transmission assets of SEBs were unbundled and transferred to corporations owned by state governments.

'Prosumers' (producing consumers) are a class of electricity suppliers not included in the list of suppliers yet. They generate renewable electricity on rooftops (see *Sun Through the Roof*, by Sunil Deambi and Shirish Garud, in the present series 'Books for the Concerned Citizen') or on farmland for their own consumption but feed the surplus into the grid. At present, grid supply for small domestic users is heavily subsidized and therefore they have insufficient incentives to move to self-generation.

The 'net metering and buy back' programme[26] provides an incentive by subsidising the capital cost and buys surplus energy at regulated rates after offsetting grid supply to the 'Prosumer'. A faster roll-out of private energy services to manage the technology challenge of self-generation by providing reasonably priced annual maintenance or 'electricity as a service' contracts, would induce a faster transition.

Has competition in supply increased?

Competition for the market has increased with unbundling and private generation. There are multiple generators but most supply is pre-contracted through long term power purchase agreements. There could be multiple discoms in a state. But retail consumers are still tied to a particular discom. However, even though competition in the market is low, benchmark competition, across discoms, helps regulators in nudging efficiency improvements. The number of licensees – admittedly a crude measure of competition – increased more than five-folds, from 61 in 1996/67 to 339 in 2019/20 (Table 11).

Table **11** Impact of competition: 1996/97 to 2019/20

Year	Supply/ generation (**GWh**)	Licensees (**number**)	Share per licensee		Average business real (**Rs billion**)
			Average supply (**GWh**)	Average cost real (**Rs/kWh**)	
1996/97	395,889	61	6490	6.35	41
2019/20	1,383,417	339	4081	5.21	21

SOURCES

Cost of power: 1996/97 – Planning Commission Annual Report on Performance of SEBs

2019/20 – PFC Performance of DISCOMS 2020–21

Other Data – CEA General Review 2020–21

Average cost and average business at real prices deflator 2015 = 100

Has the unbundling splintered supply to unviable levels?

Table 11 illustrates that as the number of suppliers increased (a crude measure of de-concentration) the average supply per licensee declined by approximately 37% as did the average business per licensee, from Rs 41 billion in 1996/97 to Rs 21 billion in 2019/20, a decline of 50% at constant (real) prices. Nevertheless, unbundling is not the reason for the financial unviability in operations.

First, consider that in the United States electricity supply of 4287 TWh (terawatt hours)[27] is shared among 2938 utilities.[28] The average supply per utility is 1459 GWh, which, if estimated at 11.1 cents per unit (kWh), amounts to an average annual business of Rs 12 billion, far lower than the Rs 21 billion (in 2015 price levels), which is the average business per licensee in India. This indicates that the low average supply volume and business revenue are not the key factors for the fiscal distress.

Second, an annual revenue of Rs 21 billion in India compares well with the revenues of companies ranked 400 to 450 out of about 5000 listed on the Bombay Stock Exchange, with their shares being actively traded depending on the company's profitability and growth outlook.

In sum the volume of sales available for market participants cannot be blamed for their low financial viability.

Measures to enhance competition
Power trade in exchanges

The ability to trade surplus power transparently is a powerful incentive for competition. An efficient discom can trade the power it has saved as can an efficient generator and enhance profits or reduce its losses. Meanwhile, strict penalties on discoms and generators for deviating from the pre-notified schedules of electricity purchase and sale ensure stable supply to customers.

Trade in renewable energy certificates

The purchase and sale of RECs (renewable electricity certificates) on the exchanges is an approximate method for assessing the proxy market price of CO_2 emissions. These certificates accrue to generators of renewable energy in proportions determined by the union government and constitute the supply side of the market.

The demand side for the RECs is created through mandates imposed by the state governments and state electricity regulators on each discom in the form of obligatory purchase of renewable energy, or the renewable purchase obligation. This obligation can be discharged by the discoms either by directly buying the stipulated quantum of electricity generated through renewable sources of energy or indirectly through buying the equivalent number of RECs on the power exchange from the holders of such certificates. Only one trade is allowed per certificate to limit speculation. The holders are not obliged to sell the certificates immediately and can wait for a better price.

The trading of RECs has imparted a degree of market orientation to profitability although it remains a thin market because of poor enforcement of the obligation by state-level regulators, who tend to prioritize near-term financial health of the discoms they regulate rather than national priorities including reducing carbon emissions.

Open access for retail customers

The ability of retail users to buy directly from any generator or a trader or a retail electricity supplier is the final stage of competition in the market. Despite legislative provisions enabling such open access, India is yet to reach that stage. Customers who have a large enough load can apply for high-tension supply for which the tariff is lower than that for retail tariff but still much more than the cost of supply. The discoms do not support open access. They fear that it may erode their profits. Consequently,

SERCs impose very heavy charges on customers opting for open access, rendering the option uneconomical.

Open access will be possible only if tariffs are restructured to reflect the actual costs of supply, with the subsidy, if any, paid directly to deserving customers. However, such payment is not easy to implement. Usually, electricity connections remain in the name of the owner of a building or a flat. A direct transfer of subsidy would benefit the owner and not the tenant, particularly, in the case of migrant workers who rent on informal basis and have low political capital.[29]

Power purchase agreements: the secret sauce that keeps turbines turning

Long-term PPAs (power purchase agreements) inhibit competition in supply but they keep the turbines running. In India the demand side of the power market, which was represented by 66 licensed discoms in 2019/20, is not financially viable.

Gujarat remains an exception where all six – four public-sector and two private-sector (Surat and Ahmedabad) – discoms are profitable. Of the nine private discoms across India,[30] seven are profitable and account for about 80% of the bulk supply bought by private discoms. The problem is that the share of the private sector in total purchase of bulk power for distribution is only 5%.

Public-sector discoms and electricity departments in Arunachal Pradesh, Goa, Mizoram, Nagaland, Puducherry, Sikkim, and Tripura, and in the union territories, buy 95% of the bulk power. Sadly, 80% of the bulk purchases are made by non-profitable publicly owned discoms. The financial condition of the electricity sector is so parlous that additional generation capacity cannot be added without 'take or pay' agreements with discoms to make the project bankable on the back of an implicit quasi-sovereign guarantee by the respective state governments.

Solar Energy Corporation of India: De-risking private investments through intermediation

Consider the financial arrangements devised to shield the generation projects from risks, which underpin the rapid increase in renewable electricity capacity. The Solar Energy Corporation of India Ltd (SECI), owned by the union government and mandated as the implementing agency for development of solar projects under the National Solar Mission, floats tenders for selection of project developers, which are selected based on tariff-based competitive bidding, SECI signs a 25-year PPA for procurement of power from these project developers. The power thus procured is in turn sold to discoms, through 25-year PPAs. In the financial year 2020/21, SECI tendered 5355 MW and awarded 12,270 MW of solar power projects.[31]

In the event of discoms defaulting on their payment, the Reserve Bank of India undertakes to deduct the outstanding dues of discoms from the central government funds to be transferred to the related state government.

SECI becoming an intermediary provides comfort to discoms, project developers, and the financing banks, because it is tacit government support in case of future disputes on commercial terms or non-payment. This arrangement is sensible in the context of the underlying financial instability of discoms but imposes implicit risks, which will grow, on the state government's budget as the ambitious plans for capacity addition in renewable electricity are realized.

Markets cannot function without creditworthy bulk buyers

Sans a stable and financially healthy distribution and retail supply sector, the proportion of merchant power – generators who voluntarily forgo the commercial insurance of long-term power purchase commitments provided by discoms – will remain limited. Merchant power generators target supply during peak times when prices are high and configure their generation capacity technology accordingly. Nuclear power, for instance, is a technology only suitable for meeting a predictable, steady demand called a 'base load'. If most of the generation is already contracted on a long-term basis by discoms, the daily sale and purchase of power on the exchanges will remain very shallow providing little business potential for merchant generators on the supply side or for demand aggregators on the demand side. Demand aggregators are licensed retail suppliers who contract to serve retail customers at agreed tariffs but who buy power from the market. The profits come from offering multiple plans, including payment plans suiting the customers profile, as in the telecom sector at present.

Breaking the monopoly of discoms over retail supply in their specified areas requires a rebalancing of the composition of customers. Some creditworthy retail customers might wish to leave discoms and sign up with independent retail suppliers. Some large-scale industrial buyers, on the other hand, might want to return to discoms if competitive tariffs are offered by them. This can happen only when the burden of cross subsidy – charging more from industry to compensate for the supply of power at less than cost to low-consumption households and farmers – is significantly diluted. Tariff should reflect the varying costs of supply at different times and voltages of supply. Until these prerequisites for competition are met, the burden of making the sector financially sound will fall disproportionately on regulators.

Autonomy for regulators

Autonomous regulatory arrangements, conceived under the Electricity Regulatory Commissions Act, 1998, and continued under the Electricity Act, 2003, have a history of only three decades. The state never conceived these entities to be autonomous in either the nature of the decisions they are authorized to take, or the extent of regulatory innovation expected of them. Although both the acts under which the commissions were conceived are less prescriptive than the Electricity (Supply) Act, 1948, and provide greater latitude in determining tariff per the regulations, regulators remain hemmed in by government policy and informal influence at the national and state levels.

At the commencement of autonomous regulation, the task before the CERC (Central Electricity Regulatory Commission) was less challenging than that before the SERCs. The national-level electricity companies were already leaders in performance benchmarks, operational metrics, and financial stability. Much of the groundwork on technical standards for capacity enhancement, performance, and monitoring had already been done by the Central Electricity Authority, although principally in the generation and transmission segments, where the central power sector companies operated.

The reform acts increased transparency and participation
The substantive change, therefore, was a shift from the relatively closed-door decision making within government – although advisory and consultative councils representing stakeholders have existed since the Indian Electricity Act, 1910 – to a more participative and transparent, quasi-judicial form of functioning, which opened what was earlier a black box to public scrutiny and debate. The structured public processes for tariff determination, an appeals process, and voluminous – and hitherto unavailable – information on quantities of supply, capital investments and asset additions, costs, contracts, business financials, and operational efficiencies shed much light on the internal functioning of discoms and enabled benchmark comparisons across companies and states.

An added advantage was the association of a more diverse set of professionals for managing the electricity sector than was possible under state control. Finance professionals, lawyers, economists, and management and development professionals were inducted into the regulatory commissions as members and staff, who brought with them much needed new perspectives and professional skills.

The decentralized structure of state-level regulators afforded a direct and more informed connection between the regulator, electricity managers and consumers. Maharashtra became fertile ground for the association of citizen groups and non-governmental organizations leveraging the efficiency of regulation. Some prominent examples are:

- Prayas in Pune: an authority on citizen-centric electricity regulation and development.
- The Centre for Science and Environment in New Delhi: active in the areas of water, energy, and citizen-centric development.
- The Energy and Resources Institute in New Delhi: a leading energy and environment research entity with hands-on experience in supporting electricity regulation.
- Consumer Unity & Trust Society in Jaipur: a public interest protagonist on competition issues.

In addition, interaction between regulators and the academic community – the IITs and IIMs – gained traction, thereby widening the pool of electricity supply researchers and professionals.

Customer grievance redress through in-house ombudsmen, within discoms, usually retired judges, were instituted by the Electricity Act, 2003 providing an easier alternative for customers than the more cumbersome civil action suits under the Consumers Protection Act.

The functioning of the SERCs remains a matter of the greatest dissatisfaction. Regulators are perceived to be too prone to second guessing government reactions in taking hard decisions like raising tariffs or passing on to consumers reasonable operational expenses through tariffs or reducing cross subsidies or not penalizing discoms for departing from supply standards. What is more common, but just as deleterious, is their patchy ability to address unjustified expenses due to high line losses, poor treasury management, or bloated capital investment in government discoms. Section 24 of the Electricity Act, 2003 empowers SERCs to revoke licenses if licensees are non-compliant or nudge them to sell their assets or both. Odisha is the only state where such directed changes in discom management have come about, partly because the licensed suppliers were private entities, and one voluntarily surrendered its license (Box 2).

State regulatory commissions crippled by misaligned public, private, and party interests

To be fair to regulators, the regulatory ecosystem, developed over the past two decades, encourages only mild and incremental changes. Choosing to live within this ecosystem means that regulators can act independently only at the margin. The legislation is structured in favour of the state through its power to issue directions to the regulator under Section 107 (CERC) and Section 108 (SERCs) of the Electricity Act, 2003. A habitually non-cooperative regulator can lose all public credibility by being summoned routinely to receive directions. Budgets of the commissions are also approved by the government, so a co-operative attitude by regulators becomes inevitable for survival.

The autonomy of a regulator at present depends on the will of the government. This situation is not unique to electricity. All regulators in India – in infrastructure or in the financial space – must find a reasonable balance between revealed public interest and the political compulsions of governments. Theoretically, there should be no daylight between the two. In reality, the compulsions of politics blur the clear line between private, public, and party interests.

The underpinnings of savvy regulators

Despite these political constraints, a few savvy regulators find space for reforms within the government's policy agenda. They strive to push reforms that align with government policy and yet advance the demand for change in other areas in the hope that their successors would be able to exploit the potential for change when the time is right.

A multi-state political economy study in India found that nurturing a culture of demand for good-quality access to electricity, combined with deft managerial improvements in the functioning of discoms, could develop a counter narrative to dilute the lazy political fallback practice of competitive politics around the price of electricity. Examples are Gujarat (2002–16), Bihar (2013–17), West Bengal (2005–2010), Delhi (2006–13), and Andhra Pradesh (1996–2004).[32] The study also perceptively red-flags extending electricity access at low prices—a fail-safe short-term political strategy but fiscally unsustainable. Escape from the resultant downward spiral needs political skill to transform the narrative, at some point, from price into quality of supply because low prices result in poor quality. Rapid economic growth helps the most. As incomes increase, the affordability of electricity improves, resulting in higher demand and the capacity to pay. Hence, energy efficiency is the key to avoid the future wasteful use of energy as India reaches higher per-capita income levels.

Nevertheless, creating more elbow room for regulators to do their job without political compulsions is a useful item on the agenda for legislative reform.

Legislative tweaks to contain populism

It is difficult to imagine the regulators being free of policy prescriptions of the union government because union law can overrule state legislation. However, the following options could be explored to reduce state influence and to extend the ambit of unhindered decision-making by regulators while remaining within the ambit of government policy.

Multi-state regulatory commissions
Convert the existing SERCs into multi-state electricity regulatory commissions to blur the artificial limitation of state boundaries

imposed on regulatory action in a networked industry like electricity. It would also serve to dilute the formal and informal influence state governments bring to bear on SERCs. Similarly, embedding a member of the CERC into such multi-state commissions will help them bypass or overcome the political economy constraints.

Multi-state representation on SERCs has a technical rationale. First, with a gradual progression to open access, much of the regulatory work will be with respect to power trading, transmission, and open access rather than tariff setting. State-level and regional management of the grid needs to improve for stable and secure operations. This requires simultaneous actions by all states on a regional grid. Such actions could include proportionate sharing of the costs of better grid management for more efficient supply. Cross-state representations in SERCs will help the evolution of a nuanced response to such changes.

Rotational membership of SERC members in the central commission

To ensure fairness and harmony between the union and the state governments, consider the appointment of two representatives from state governments on rotation. The Forum of Regulators, formalized under Section 166(c) of the Electricity Act, 2003, could be entrusted with the task of rotating the membership amongst the SERCs. The forum did a remarkable job in coordinating the policy for determining the proportion of renewable purchase obligations state by state, which demonstrated the potential for collaborative decision-making.

Such mechanisms for joint decision-making between the union and states would align with the spirit of cooperative federalism. It could enhance collegiality between the members of the CERC and SERCs in the same manner as the All-India Services purportedly do between the staff of the union and state governments.

Reporting to the Parliament or the state legislatures
Make regulators directly answerable to empowered, specially constituted, multiparty subcommittees of the Parliament (for CERC) and of the respective state legislatures (for SERCs) which would also endorse their appointment on selection by the respective government. If these subcommittees are designed to represent the share of electoral votes rather than the number of seats won in the Parliament or the state legislative assembly, this procedural tweak could broaden political engagement with regulators and protect them from undue influence by the ruling parties.

Empower SERCs to enforce fiscal discipline on subsidies
Deduct state-level power subsidies not paid in time to discoms from central funds and pay the subsidies directly to bulk electricity supplying entities on the recommendation of the local regulator. Here the state-level regulators would be 'trusted advisors' of the union government. This measure would add a useful constraint to state-level populist excesses and raise the profile of SERCs to act independently, although it might transgress state sovereignty.

The institutional changes described above require states to give up some authority to bordering states, on a mutual basis – a politically sensitive issue which requires a great deal of political capital to implement. They also require a spirit of accommodation between the states and the union government, which again implies that political capital would need to be expended.

Why bother then? Presently regulators tend to play a subordinate role to government. Once a regulatory commission is constituted, the state must vacate the policy space and allow the regulator to lead on policy issues based on research and wide prior consultation, subject to ratification by the state. Unless regulatory bodies are known and seen to act on technical grounds, much of the efficiency enhancement expected from those bodies will fail to be realized, to the detriment of competition, economic growth, and consumer benefit.

The needs of viable power markets

Once the demand side of the market (discoms) is financially solvent and bulk electricity suppliers have the option of either entering into long-term contracts with retail users and retail demand aggregators or to access the power market, variation and volatility could increase. Price volatility in supply can increase due to the greater share of variable renewable electricity traded on the market in the future.

A volatile and variable electricity market can destabilize grid management unless there is sufficient 'quick ramping up' of contracted surplus capacity (including storage capacity) to fill the gaps in variable supply. Presently, grid supply standards are maintained by states suppressing demand to avoid buying expensive power on the spot market, resulting in enforced outages of supply (load shedding). This practice lowers the operational efficiency and quality of grid supply.

Renewable capacity is targeted to increase from 25% to 50% by 2030, and the quality and price of support services – battery storage, gas turbine support or stored hydro – will become the key determinants of the price of delivered retail electricity. As for the energy generated, 1 MW of renewable energy capacity equals less than 0.5 MW of fossil fuel capacity because of the intermittent and variable availability—limited to the daytime for solar power and to seasonal winds for wind power. The higher the proportion of variable power, the stronger the need for spinning reserves,[33] operating reserves and ancillary capacity available with POSOCO, the system operator, to manage the variable supply.

At present, the generating stations not likely to be 'dispatched' on any day, owing to their higher cost, are designated to provide spinning reserve, and are compensated for the consequential costs on heat loss, fuel use, etc.

System operations in the national, regional, and state grids must be aligned for effective grid management. The NLDC (National Load Dispatch Centre), RLDCs (regional load dispatch centres), and SLDCs (state load dispatch centres) must have adequate autonomy and financial and technical resources to call on or contract standby ancillary capacity to manage grid frequency or maintain grid stability.

The need for a 'smart' grid

One area that deserves attention is whether the system operators should be penalized for grid instability or collapse. The flip side of fixing responsibility is providing the means for proactive, computerized system controls, which cannot be overridden by generators or discoms. Investing in integrated control systems for a smart grid, which generates red flags well in advance of the occurrence of a fault based on artificial intelligence and machine learning, needs to be implemented not only in the national and regional grids but also in the state grids.

The most basic requirement of smart system control is the pervasive metering of electrical flows right up to the end-user's meter. This is still a work in progress. Apparently, inadequate domestic capacity for manufacturing meters is a constraint.[34] The recent controls and duties on import could adversely affect the supply of meters.

Four short-term hard choices
Pay or wait to go green

Once the economy revives, demand will increase, and electricity will once again be in short supply. Choosing between reviving stranded fossil fuel assets to meet the increased demand or facing the unemployment and economic dislocation from their closure and the higher cost of natural gas or nuclear energy will be a painful choice. A hard 'no return to conventional fossil fuels policy' becomes expensive—as Germany now realizes. It decided to shut down nuclear generation following the Fukushima (Japan) nuclear power disaster and now must grapple with the explosion in gas prices caused by the Ukraine crisis.

Prioritise electricity supply and growth with least-cost carbon mitigation

India needs a medium-term strategy till 2040 for filling the energy gap when the sun is not shining, and the wind is not blowing in the medium term. The availability of cost-effective storage is uncertain. Continuing to use depreciated coal plants does not align with carbon mitigation objectives unless carbon capture measures are enforced. Administered cutbacks on electricity consumption, to balance the grid during the intermittent periods when renewable energy is unavailable is counterproductive for well-being, economic productivity, and climate goals. The cost of power sector distortions was estimated at 4.1% of GDP, or $86.1 billion in 2015/16,[35] which is more than the growth rate in the last quarter of 2019/20 and equal to the expected growth in the last quarter of 2022/23.

Choose between alternative paths for green power

Until recently, natural gas was the favoured transition fuel, with its fast ramp-up and ramp-down capability and low carbon emissions. But the Ukraine situation has shaken the credibility of LNG (liquefied natural gas) as a reasonably priced source of electricity. If world trade in general reduces and the 'open economy miracle' diminishes into fractured markets including for oil and gas, long term LNG supply will need to be contracted at viable levels. Presently India is unable to use fully the available 25 GW of gas capacity because of fuel price concerns.

Optional renewable fuels such as ethanol have the downside of water intensity, which a water-scarce economy like India can ill afford on large scale, beyond the additional income it offers to sugar cane farmers. The proposal to use starchy cereals like rice or maize faces the same constraints. Agriculture consumes 78% of the water available in India and still less than half of the cropped area is irrigated. Enhancing this share seems unrealistic. The existing challenge is to reduce the share without affecting productivity through scientific water management, but this needs pricing reform and change in traditional farming practices.

Continuing use of coal, oil, and gas by adding technology to carbon capture, use, and store (CCUS) could make a better transition option especially since we have till 2070 (nearly half a century) to reach net zero emissions. Adoption of CCUS technology at scale could reduce the cost to reasonable levels.

Traditionally gas capacity is favoured for grid support, but using renewable energy combined with storage (pumped or battery) could be another option, which make renewable energy a part of grid support on its commercial merits – on the same terms as energy generated from fossil fuels – instead of its mandated, privileged 'must run' status.

Choose between protecting domestic industry and allowing cheaper imports

The price of solar electricity decreased by more than 80%, from Rs 10.95/kWh to Rs 1.99/kWh over 2010–2020 making solar power a win-win option. However, the subsequent imposition of high import duties, restriction on imports limiting them only to authorized manufacturers (mostly domestic), and the rise in the prices of commodities in February 2022 post the Ukraine crisis are likely to check the continuing decline in the price of solar electricity, which is likely to stabilize instead. India can ill afford to reverse its open-economy stance and revert to the pre-1991 industrial policy protecting domestic manufacturing from the competition offered by imports. How will we escape the obvious downsides of protection-driven domestic manufacturing fostering complacency and inefficiency?

Trends favouring India

India has wisely chosen 2070 as the target for lowering its carbon emissions to a level which equals domestic absorption capacity (achieving the 'net zero' status). The intervening decades are likely to see four major changes to facilitate that transition.

High growth can make green power affordable

Economic growth can progressively increase household income and generate additional revenues for the state to make clean energy more affordable and its use in transport, industry, and homes more prevalent.

Renewable and hydroelectricity offer unutilized potential

The global push for electrification favours India because the country is well endowed with unused renewable electricity potential from hydro, solar, and wind. The promise of green hydrogen emerging as a commercially viable option for industrial use, as a blend in natural gas, and as a medium for energy storage beyond 2030 evokes hope that the energy transition can be managed in time and at reasonable cost.

Growth of digital connectivity

India is on the fast track to being a digital economy, which is the global pathway, facilitated by electricity, to future growth, well-being, and security. The provisioning for 5G services, which is underway, could induce the third significant transformation of the Indian economy, unleashing economic productivity, enhancing the effectiveness of public services and deep social transformation. This is why upgrading the electricity sector is a necessary condition for sustainable growth.

Symmetric policy preferences across parties

India is wedded to continuity with change – a dogma that slows down the process of reform. The good news is that, consequently, there is a broad consistency across the economic strategies of the successive governments, including those which managed India since 1950. There have been periodic disruptions. The period of industrial and financial sector nationalisation in the 1970s was one such. Two decades later the 1991 liberalization and reforms were also a point of departure which boosted economic growth. However, follow-through reforms became hostage to political expediency slowing growth to predictable levels. India rarely strays too far from the mean, making the trend line predictable, sometimes frustratingly so. The good news is that the long-term trend in growth has been positive since 1950.

The third decade of this century has dawned with reminders that India needs to strengthen its energy resilience and security. This is possible if the spirit of cooperative federalism suffuses institutional arrangements and if public interest trumps political upmanship. Less government, more governance, and a growing share in the economic pie for the private sector, including in electricity supply, remain sensible mantras to implement.

Bibliography and notes

1. Tirthankar Roy. *The Economic History of India*. Oxford University Press. 2000.
2. https://www.britannica.com/technology/railroad/The-Liverpool-and-Manchester-Railway
3. https://www.britannica.com/technology/telegraph
4. For a deep but simply described dive into the science of electricity there can be no better reference than Chapter 2 of *Know Your Power – A citizens' primer on the electricity sector* published by Prayas Energy Group, Pune in 2004 (revised edition 2019).
5. https://www.americaslibrary.gov/jb/gilded/jb_gilded_hydro_3.html
6. Siddhartha Chandra Mukherjee. "Steps of power: Notes on the history of electrification in India (1883–1930)". Proceedings of the Indian History Congress, Vol. 78. 2017. pp. 498–506. https://www.jstor.org/stable/26906120. Accessed 1 May 2022.
7. Central Electricity Authority. General Review. 2019–20.
8. N Govinda Rao, Kalirajan K P, Shand R. *The Economics of Electricity Supply in India*. Macmillan India Limited. 1998.
9. Economic issues in the power sector. World Bank Country Study, PUB2335. November 1979.
10. *The Economic Weekly*. 20 August 1960.
11. Industrial Policy Resolution. Government of India. 1956.
12. Coase R H. *Land Economics* Vol. XXVI No. 1. "The nationalization of electricity supply in Great Britain." February 1950.
13. Gillion Simmonds. *Regulation of the UK Electricity Industry*. University of Bath. 2002.
14. Transmission at higher voltage is cost-effective since it reduces the 'line loss' and can transmit higher volumes of energy, although it needs thicker conducting cables and associated transformation equipment.

15 Industrial Policy Statement. Department of Industrial Development, GOI. 24 July 1991.
16 https://powermin.gov.in/en/content/policy-setting-mega-power-projects-pvt-sector
17 https://powermin.gov.in/en/content/policy-setting-mega-power-projects-pvt-sector
18 Energy shortage is calculated as the difference between supply and demand in any given period. Peak shortage relates to specific and predictable periods during the day or seasons when demand surges for a few hours before subsiding. In India there is a morning peak when people wake up, start using electricity to cook or get ready for work and again in the evening when people return home from work. Traditionally seasonal peak happens during the hot weather while seasonal lows happen during the monsoons when agricultural demand is low. When to use the washing machine or your dish washer is one such case where customers can help in grid management and save on electricity bills by shifting demand to off-peak times. Once 5G connectivity is available and the Internet of Things (IoT) is implemented, options for managing your electricity bill, using electricity more efficiently, even without drastic behaviour change, will multiply.
19 *Himal Sanchar.* 2 November 2021.
20 Report on performance of power utilities. Power Finance Corporation. 2019–20.
21 Vijay Joshi. *India's Long Road: The Search for Prosperity.* Penguin. 2016.
22 Isher Judge Ahluwalia. *Industrial Growth in India: Stagnation Since the Mid-sixties.* Oxford University Press. 1985.
23 David C. Engerman. *The Price of Aid.* Harvard University Press. 2018.
24 Vijay Joshi. *India's Long Road: The Search for Prosperity.* Penguin. 2016

25 Central Electricity Authority. General Review. 2019–20.
26 The PM-KUSUM (Kisan Urja Suraksha evam Utthan Mahaabhiyan) supports farmers to set up solar power plants with capacity between 500 kW and 2 MW with an assured buy back by the discom at the lower rate of Rs 0.40/kWh or Rs 0.65 million per MW/per year. Also, diesel water pumps or grid-connected electric pumps can be converted to solar powered pumps with a subsidy amounting to 60% of the cost (shared equally between the union and the state governments), loan from a bank amounting to 30% of the cost, and an upfront investment of only the remaining 10% by the farmer.
27 One Terawatt hour (TWh) equals 1000 Gigawatt hours (GWh).
28 https://www.eia.gov/todayinenergy/detail.php?id=40913.
29 Prayas (Energy Group). Comments and suggestions on Draft National Electricity Policy. 13 May 2021.
30 Three in Delhi (BSES Rajdhani Power Limited, BSES Yamuna Power Limited, and Tata Power Delhi Distribution Limited), two in Gujarat (Torrent Power, Surat and Ahmedabad), one in Maharashtra (Adani Electricity Mumbai Limited), in Uttar Pradesh, Noida Power Company Limited, and two in West Bengal (Calcutta Electricity Supply Company and India Power Corporation Limited).
31 Solar Energy Corporation of India Limited. Annual Report, 2020–21. 2021.
32 Dubash N K, Kale, Sunila S, Bharvikar R (eds). *Mapping Power: The Political Economy of Electricity in India's States*. Oxford University Press. 2018.
33 Spinning reserve is generating capacity equal to 5% of the demand load kept in operation on part load (like a stationary car with a running engine), synchronized with the grid and ready to provide increased generation in response to a frequency dip. This is divided into a primary response of 4 GW at the national level, where generation follows frequency

fluctuations stabilizing it – much like the stabilizer in your home controls the voltage fluctuations. Secondary reserve is maintained in the regional grids, equal to the size of the largest generating unit. This kicks in if the primary reserve is unable to stabilize the grid near instantly. At the state level, tertiary reserves are maintained equal to 50% of the size of the largest generating unit which kick in if grid instability continues for a few minutes. Deo P, Chatterjee S K, Modak S. *Renewable Energy in India: Economics and Market Dynamics.* Sage. 2021. pp. 164–165.

34 Deo P, Chatterjee S K, Modak S. *Renewable Energy in India: Economics and Market Dynamics.* Sage. 2021.

35 Zhang Fan. *In the Dark: How Much Do Power Sector Distortions Cost South Asia?* World Bank. 2019.